Financial Systems, Corporate Investment in Innovation, and Venture Capital

Financial Systems, Corporate Investment in Innovation, and Venture Capital

Edited by

Anthony Bartzokas

Professor of Development Economics, University of Athens and Researcher, United Nations University–Institute for New Technologies, Maastricht, The Netherlands

and

Sunil Mani

Researcher, United Nations University–Institute for New Technologies, Maastricht, The Netherlands

Edward Elgar

Cheltenham, UK • Northampton, MA, USA

Published by
Edward Elgar Publishing Limited
Glensanda House
Montpellier Parade
Cheltenham
Glos GL50 1UA
UK

Edward Elgar Publishing, Inc.
136 West Street
Suite 202
Northampton
Massachusetts 01060
USA

A catalogue record for this book
is available from the British Library

Library of Congress Cataloguing in Publication Data
Financial systems, corporate investment in innovation, and venture capital /
edited by Anthony Bartzokas and Sunil Mani.
p. cm.
Includes bibliographical references.
1. Venture capital. 2. High technology industries—Finance. I. Bartzokas, Anthony, 1962– II. Mani, Sunil.

HG4751.F55 2004
332′.04154—dc22 2003064350

ISBN 1 84376 392 3

Printed and bound in Great Britain by MPG Books Ltd, Bodmin, Cornwall

Contents

Preface

This book draws on earlier work on the role of innovation policies and investment decisions in developing countries conducted by the two editing authors in the last five years. In a number of country case studies and in comparative analysis of patterns of industrialization, we reached the conclusion that, in the current context of world markets and innovation-driven competition, a small number of advancing developing countries have introduced policy instruments which facilitated sustainable and profitable business initiatives in the local economy. This has been an evolving process with many unique country-specific characteristics, drawbacks and adjustments to external factors and exogenous shocks. Our emphasis on detailed empirical analysis led us to the conclusion that patterns of investment decisions and access to finance have been important driving forces in these processes.

This volume is our first step towards a more systematic analysis of this hypothesis. Our emphasis is on the role of venture capital as a core element of the institutional framework supporting innovation dynamics in developing countries. Our work was organized on three planes: an extensive review of literature and available data on the role of venture capital in developing countries with specific emphasis on Asian economies, detailed case studies prepared by country experts and additional invited contributions on the linkages between finance and innovation. A conference organized in Brussels on 6 and 7 November 2002, with financial support from the European Commission, facilitated our work and provided a valuable opportunity for a detailed discussion of our research findings with leading experts. We are grateful to all the colleagues who attended that meeting for their comments and suggestions. Our thanks to Andrew Sors and Nikos Kastrinos from DG Research who supported and encouraged our work. At UNU–INTECH Eveline in de Braek and Monique Seuren provided excellent secretarial support and Ad Notten responded promptly to our long lists of requests for specialized material on venture capital. Of course, the responsibility for the views and any mistakes in this volume remains with us.

A. Bartzokas and S. Mani
July 2003, Maastricht

Contributors

Gil Avnimelech, *School of Business Administration, Hebrew University, Jerusalem*

Anthony Bartzokas, *University of Athens and United Nations University–Institute for New Technologies, Maastricht*

B. Bowonder, *Administrative Staff College of India, Hyderabad*

Jian Gao, *School of Economics and Management, Tsinghua University, Beijing*

Bronwyn Hall, *Department of Economics, University of California, Berkeley*

Kyonghee Han, *Institute for Social Development Studies, Yonsei University, Seoul, Korea*

Martin Kenney, *Department of Human and Community Development, University of California, Davis*

Sunil Mani, *United Nations University–Institute for New Technologies, Maastricht*

Colin Mayer, *Said Business School and Wadham College, University of Oxford*

László Szerb, *Faculty of Business and Economics, University of Pécs, Hungary*

Shoko Tanaka, *Department of Human and Community Development, University of California, Davis*

Morris Teubal, *Department of Economics, Hebrew University, Jerusalem*

Attila Varga, *Faculty of Business and Economics, University of Pécs, Hungary*

Clement Wang, *Centre for Entrepreneurship, National University of Singapore*

Steven White, *INSEAD, Paris*

Wei Zhang, *School of Economics and Management, Tsinghua University, Beijing*

1. Introduction

Anthony Bartzokas and Sunil Mani

The conditions for successful manufacturing have changed considerably in recent years. The allocation of industrial investment in plant, equipment and intangibles is changing markedly and new products and processes encompass a greater number of different technologies. These trends involve different investment needs. The main distinction is between 'core technological investment' (comprising the bulk of technological investment) and 'complementary investment' (which guarantees the efficiency of the 'core investment'). Another dimension is the distinction between tangible and intangible investment, with the understanding that the latter comprises the bulk of technological investment. The decision to invest in new technologies is constrained by uncertainty and information costs. Uncertainty is particularly high when technologies are new and still changing rapidly and investments are considerable.

Because liquidity risk is positively related to firm size and because barriers to credit increase the risk of doing business, entrepreneurs unable to insure themselves against large risks may prefer to remain small and to diversify their activities in whatever way they can. For instance they may start a new firm instead of expanding the one they currently operate. Barriers to credit also affect technology choices. If access to credit is partly determined by the collateral value of the investment, purchases of land, buildings and vehicles are facilitated while the building up of stocks, wage fund and credit to customers are not. This may result in the adoption by large firms of capital-intensive methods of production and in an emphasis on production instead of marketing and product improvement, even though labour-intensive methods may be more efficient and an improvement in marketing much needed. Moreover, in areas of fast-changing technologies, small firms and start-ups may be unable to afford the best available technique of production. Risk also makes firms reluctant to experiment with the unknown. They may reduce risk by opting for a flexible organization of their business. For instance, investors may prefer multi-purpose technologies that can easily be applied to new tasks, even if it means bypassing state-of-the-art specialized equipment. They may avoid investments in equipment and technology not because they could not get a bank to finance it but because rigid loan repayment obligations would put the firm at risk. In all these cases, some of the gains from specialization and learning by doing are not captured and the size distribution of firms remains inefficient.

A large body of empirical literature has investigated the impact of information problems in financial markets on investment decisions of firms in developing countries. The increased availability of panel data has resulted in a growing volume of empirical work in recent years. Despite the intense debate, evidence on the effects of different financial systems is still sparse. The range of factors which bear on cross-firm or cross-country variations in performance is considerable. This does not stop many from equating differences in economic performance between countries with their different types of financial systems. What is more realistic than trying to provide a general typology at the aggregate level is to consider the way in which financial systems can bear important aspects of the performance of the corporate sector.

Financing of knowledge production is charecterized by at least two types of failures. The first has been very well articulated in the so-called 'appropriability argument'. This argument runs as follows: R&D investments result in the production of new knowledge and this is non-rival in its use. Despite the existence of intellectual property right (IPR) mechanisms, given its non-rival nature, it can be copied or imitated by competitor firms at costs which are less than the cost of creating it from scratch. Economists have attempted to capture this by computing the spillover gap or the gap between private and social rates of returns for samples of innovation. The existence of this gap justified various public policy measures to combat possible underinvestment in R&D by private sector agents. These public policy measures range from various fiscal incentives for R&D, research grants, strengthening of the IPR regime, financing of research partnerships and so on. The major assumption in this line of argument is that the firm or the agent which performs the R&D is also its financier.

The second type of failure exists when the innovation investor and the financier are two different entities. Under such circumstances a second gap exists between the private rate of return and the cost of capital. This implies that the conventional capital market, whether based on debt or on equity, would eschew projects that result in innovations as the output of these projects are uncertain or the projects are such that one cannot even attach probabilities to their potential outcomes. Hall (Chapter 2 in this volume) has identified three main types of reason for the existence of a gap between external and internal costs of capital:

1. asymmetric information between inventor and investor;
2. moral hazard on the part of the investor, or arising from the separation of ownership and management; and

3. tax considerations that drive a wedge between external finance and finance by retained earnings.

The response to this has come from the private sector itself but very often supported by state funds, namely the establishment of specialized financial agencies such as venture capital institutions. In short knowledge production is characterised by two types of market failure and state intervention is required to offset for the consequent shortfalls in investment. Figure 1.1 summarizes this point.

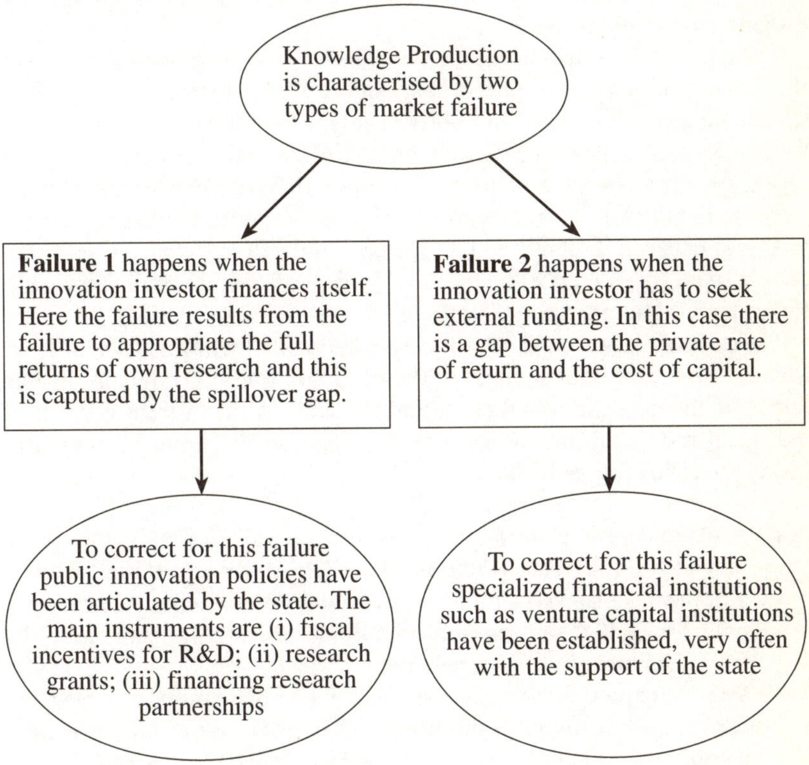

Source: Own compilation.

Figure 1.1 Rationale for state intervention in knowledge generation

Venture capital is a key component of the development of high-technology industry, and an essential element for policies to develop such industries in all countries, developed as well as developing. This book is a first attempt to analyse the link between the corporate sector and local financial systems in developing countries. The hypothesis we put forward in this book is that recent trends in the corporate sector, with increasing competition and structural adjustment and the deregulation of financial institutions in advanced developing countries, are increasingly creating new channels of interaction with profound implications for investment decisions and patterns of corporate development. So far, the innovation studies perspective has paid limited attention to these issues. It has been implicitly assumed that, in the context of a heavily regulated financial system, the provision of credit and the availability of other financial instruments are being driven by the internal dynamics of manufacturing industry and to a lesser extent by government policy.

The empirical contributions included in this volume reconsider some of these assumptions with a proper mapping of technological and organizational changes in the two main poles of interaction, that is, financial institutions and industry in the current context of industrial restructuring and services deregulation. In chapter 6, Mani and Bartzokas highlight these issues in detail. A first set of contributors present analytical essays on the linkages between finance and corporate innovation, while a series of country case studies report on the development of venture capital in China, India, Israel, Hungary and Singapore.

The country case studies have focused on careful mapping of technological and organizational changes in the two main poles of interaction: new forms of financial intermediation and industry in the current context of industrial restructuring and services deregulation. In doing so, they produced the following insights:

- Venture capital is an important source of institutional support for new-technology-based ventures especially in their early stage. Unless emphasis is placed on financing technology-based small and medium enterprises at their seed and start-up stages, venture capital will be less of an incentive to innovation.
- Venture capital firms require strong public policy support in terms of tax and other financial incentives. They also require proper exiting mechanisms such as a well developed stock market and an adequate supply of well trained professionals especially at the 'due diligence' stage.
- A distinguishing feature of venture capital finance is the additional support and knowledge transfer provided by the investing firms to the firm they invest in.

In addition to these general points the country case studies demonstrate that (a) there is increasing importance of government backing for the setting-up of venture capital industries, virtually in all countries with an innovation policy; and (b) there is a trend towards globalization of venture capital, through the incorporation of venture capital funds in global portfolio management structures.

Barriers to credit generate allocative inefficiencies and pull resources away from manufacturing. These problems are being generated because the investment projects that are financed may not be those with the highest return. This is true whenever there is not a perfect match between investment opportunities and the allocation of credit. If firms that have long been in existence find it easier to gain access to credit while new firms cannot do so, certain firms will outlive their usefulness and competition through firm entry will be thwarted. The sectoral allocation of investment is affected because lost investment opportunities and inefficient production choices reduce aggregate returns to industrial capital. As a result, funds are channelled to uses other than manufacturing: commerce, government bonds and capital flight. This process is reinforced if, to reduce their exposure to risk, investors are drawn towards operations with a rapid turnover, such as commerce, or to financial investments with a safe return. That the influence of global capital markets, the local circumstances and the diversity of configurations of venture capital should be the three main pillars of policy using venture capital to promote investment in high technology and innovation is one of the main conclusions of the case studies.

If financial markets are underdeveloped, people will choose poorly productive, but flexible, technologies. Given these technologies, producers do not experience much risk, and hence there is little incentive to develop financial markets. Conversely, if financial markets are developed, technology will be more specialized and risky, thereby creating the need for financial (and assets) markets. A particular resource (capital) can be specialized into a narrow range of tasks without being harmed through the increase of risk because financial institutions are used in order to deal with it. Thus financial markets contribute to growth by facilitating a greater division of labour. In the absence of financial markets, diversification is taking place at the firm level through technology 'options'. Firms will choose technologies that are less risky, with many applications, but less productive. Firms are reluctant to engage in sophisticated technologies as long as they cannot share the risk they incur with financial markets. Indeed, there is a strategic complementarity between financial markets and technology, because both are instruments that can be used for diversification and technological upgrading.

As governments invest more in venture capital, what are the challenges they face in their policy designs and evaluations? The globalization of venture capital poses questions about the ability of local policy makers to harness venture capital for the development of local technological capabilities. In that context, this volume clearly demonstrates the great diversity of venture capital structures and roles in innovation processes. For example, Singaporean venture capital and Chinese venture capital organizations are very different in structure, aspirations, financial power and ways in which they receive support from their respective governments. Both, however, invest heavily in innovative companies in mainland China. Somewhat paradoxically, globalization has increased the importance of local conditions in every attempt to attract international investment and channelling these resources into the development of sustainable businesses in the high-tech sector.

2. The financing of research and development

Bronwyn Hall[1]

INTRODUCTION

It is a widely held view that research and development (R&D) activities are difficult to finance in a freely competitive market place. Support for this view in the form of economic–theoretic modeling is not difficult to find and probably begins with the classic articles of Nelson (1959) and Arrow (1962), although the idea itself was alluded to by Schumpeter.[2] The argument goes as follows: the primary output of R&D investment is the knowledge of how to make new goods and services, and this knowledge is non-rival – use by one firm does not preclude its use by another. To the extent that knowledge cannot be kept secret, the returns to the investment in it cannot be appropriated by the firm undertaking the investment, and therefore such firms will be reluctant to invest, leading to the under provision of R&D investment in the economy.

Since the time when this argument was fully articulated by Arrow, it has of course been developed, tested, modified and extended in many ways. For example, Levin *et al.* (1987) and Mansfield *et al.* (1981) found, using survey evidence, that imitating a new invention was not costless, but could cost as much as 50–75 per cent of the cost of the original invention. This fact will mitigate but not eliminate the underinvestment problem. Empirical support for the basic point concerning the positive externalities created by research that was made by Arrow is widespread, mostly in the form of studies that document a social return to R&D that is higher than the private level (Griliches, 1992; Hall, 1996). Recently, a large number of authors led by Romer (1986) have produced models of endogenous macroeconomic growth that are built on the increasing returns principle implied by Arrow's argument that one person's use of knowledge does not diminish its utility to another (Aghion and Howitt, 1997).

First published in the *Oxford Review of Economic Policy*, 18(1).

This line of reasoning is already widely used by policy makers to justify such interventions as the intellectual property system, government support of R&D, R&D tax incentives and the encouragement of research partnerships of various kinds. In general, these incentive programmes can be warranted even when the firm or individual undertaking the research is the same as the entity that finances it. However, Arrow's influential paper also contains another argument, again one which was foreshadowed by Schumpeter and which has been addressed by subsequent researchers in economics and finance: the argument that an additional gap exists between the private rate of return and the cost of capital when the innovation investor and financier are different entities.

This chapter concerns itself with this second aspect of the market failure for R&D investment: even if problems associated with incomplete appropriability of the returns to R&D are solved using intellectual property protection, subsidies or tax incentives, it may still be difficult or costly to finance R&D using capital from sources external to the firm or entrepreneur. That is, there is often a wedge, sometimes large, between the rate of return required by an entrepreneur investing his or her own funds and that required by external investors. By this argument, unless an inventor is already wealthy, or firms already profitable, some innovations will fail to be provided purely because the cost of external capital is too high, even when they would pass the private-returns hurdle if funds were available at a 'normal' interest rate.

In the following, I begin by describing some of the unique features of R&D investment. Then I discuss the various theoretical arguments why external finance for R&D might be more expensive than internal finance, going on to review the empirical evidence on the validity of this hypothesis and the solutions that have been developed and adopted by the market and some governments. The chapter concludes with a discussion of policy options.

RESEARCH AND DEVELOPMENT AS INVESTMENT

From the perspective of investment theory, R&D has a number of characteristics that make it different from ordinary investment. First, and most importantly, in practice 50 per cent or more of R&D spending is the wages and salaries of highly educated scientists and engineers. Their efforts create an intangible asset, the firm's knowledge base, from which profits in future years will be generated. To the extent that this knowledge is 'tacit' rather than codified, it is embedded in the human capital of the firm's employees, and is therefore lost if they leave or are fired.

This fact has an important implication for the conduct of R&D investment. Because part of the resource base of the firm itself disappears when such workers leave or are fired, firms tend to smooth their R&D spending over time, in order to avoid having to lay off knowledge workers. This implies that R&D spending at the firm level typically behaves as though it has high adjustment costs (Hall *et al.*, 1986; Lach and Schankerman, 1988), with two consequences, one substantive and one that affects empirical work in this area. First, the equilibrium required rate of return to R&D may be quite high simply to cover the adjustment costs. Second, and related to the first, is that it will be difficult to measure the impact of changes in the costs of capital, because such effects can be weak in the short run owing to the sluggish response of R&D to any changes in its cost.

A second important feature of R&D investment is the degree of uncertainty associated with its output. This uncertainty tends to be greatest at the beginning of a research programme or project, which implies that an optimal R&D strategy has an options-like character and should not really be analysed in a static framework. R&D projects with small probabilities of great success in the future may be worth continuing even if they do not pass an expected-rate-of-return test. The uncertainty here can be extreme and not a simple matter of a well-specified distribution with a mean and variance. There is evidence, such as that in Scherer (1998), that the distribution of profits from innovation sometimes has a Paretian character where the variance does not exist. When this is the case, standard risk-adjustment methods will not work well.

The natural starting point for the analysis of R&D investment financing is the 'neoclassical' marginal profit condition, suitably modified to take the special features of R&D into account. Following the formulation in Hall and Van Reenen (2000), I define the user cost of R&D investment ρ as the pre-tax real rate of return on a marginal investment that is required to earn r after (corporate) tax. The firm invests to the point where the marginal product of R&D capital (MPK) equals ρ:

$$MPK = \rho = \frac{1 - A^d - A^c}{1 - \tau}(r + \delta + MAC)$$

where τ is the corporate tax rate, δ is the (economic) depreciation rate, and MAC is the marginal adjustment cost.

In this equation, A^d and A^c are the present discounted values of depreciation allowances and tax credits, respectively. In most financial accounting systems, including those used by major OECD economies, R&D is expensed as it is incurred, rather than capitalized and depreciated, which means that the lifetime of the investment for accounting purposes is much shorter than the economic life of the asset created and that A^d is simply equal to τ for tax-paying firms. Many countries have a form of tax credit for R&D, either incremental or otherwise, and this will be reflected in a positive value for A^c.[3] Note that when A^c is zero, the corporate tax rate does not enter into the marginal R&D decision, because of the full deductibility of R&D.

The user-cost formulation above directs attention to the following determinants of R&D financing:

(i) Tax treatment such as tax credits, which are clearly amenable to intervention by policy makers.
(ii) Economic depreciation δ, which in the case of R&D is more properly termed obsolescence. This quantity is sensitive to the realized rate of technical change in the industry, which is in turn determined by such things as market structure and the rate of imitation. Thus it is difficult to treat δ as an invariant parameter in this setting.
(iii) The marginal costs of adjusting the level of the R&D programme.
(iv) The investor's required rate of return r.

The last item has been the subject of considerable theoretical and empirical interest, on the part of both industrial-organization and corporate-finance economists. Two broad strands of investigation can be observed: one focuses on the role of asymmetric information and moral hazard in raising the required rate of return above that normally used for conventional investment, and the other on the requirements of different sources of financing and their differing tax treatments for the rate of return. The next section discusses these factors.

THEORETICAL BACKGROUND

This section reviews in more detail the reasons why the impact of financial considerations on the investment decision may vary with the type of investment and with the source of funds. To do this, I distinguish between those factors that arise from various kinds of market failures in this setting and the purely financial (or tax-oriented) considerations that affect the cost of different sources of funds.

One of the implications of the Modigliani–Miller theorem (1958, 1961) is that a firm choosing the optimal levels of investment should be indifferent to its capital structure, and should face the same price for investment and R&D investment on the margin. The last dollar spent on each type of investment should yield the same expected rate of return (after adjustment for non-diversifiable risk). A large literature, both theoretical and empirical, has questioned the bases for this theorem, but it remains a useful starting point.

Reasons why the theorem might fail in practice are several: (i) uncertainty coupled with incomplete markets may make a real options approach to the R&D investment decision more appropriate; (ii) the cost of capital may differ by source of funds for non-tax reasons; (iii) the cost of capital may differ by source of funds for tax reasons; and (iv) the cost of capital may also differ across types of investments (tangible and intangible) for both tax and other reasons.

With respect to R&D investment, economic theory advances many reasons why there might be a gap between the external and internal costs of capital; these can be divided into three main types:

(i) asymmetric information between inventor and investor;
(ii) moral hazard on the part of the inventor or arising from the separation of ownership and management;
(iii) tax considerations that drive a wedge between external finance and finance by retained earnings.

Asymmetric-information Problems

In the R&D setting, the asymmetric-information problem refers to the fact that an inventor frequently has better information about the likelihood of success and the nature of the contemplated innovation project than potential investors. Therefore, the marketplace for financing the development of innovative ideas looks like the 'lemons' market modelled by Akerlof (1970). The lemons' premium for R&D will be higher than that for ordinary investment because investors have more difficulty distinguishing good projects from bad when the projects are long-term R&D investments than when they are more short-term or low-risk projects (Leland and Pyle, 1977). When the level of R&D expenditure is a highly observable signal, as it is under current US and UK rules, we might expect that the lemons' problem is somewhat mitigated, but certainly not eliminated.[4]

In the most extreme version of the lemons model, the market for R&D projects may disappear entirely if the asymmetric-information problem is too great. Informal evidence suggests that some potential innovators believe this to be the case in fact. And, as is discussed below, venture-capital systems are viewed by some as a solution to this 'missing markets' problem.

Reducing information asymmetry via fuller disclosure is of limited effectiveness in this arena, owing to the ease of imitation of inventive ideas. Firms are reluctant to reveal their innovative ideas to the marketplace and the fact that there could be a substantial cost to revealing information to their competitors reduces the quality of the signal they can make about a potential project (Bhattacharya and Ritter, 1983; Anton and Yao, 1998). Thus the implication of asymmetric information coupled with the costliness of mitigating the problem is that firms and inventors will face a higher cost of external than internal capital for R&D owing to the lemons' premium.

Some empirical support for this proposition exists, mostly in the form of event studies that measure the market response to announcements of new debt or share issues. Both Alam and Walton (1995) and Zantout (1997) find higher abnormal returns to firm shares following new debt issues when the firm is more R&D-intensive. The argument is that the acquisition of new sources of financing is good news when the firm has an asymmetric-information problem because of its R&D strategy. Similarly, Szewczyk *et al.* (1996) find that investment opportunities (as proxied by Tobin's q) explain R&D-associated abnormal returns, and that these returns are higher when the firm is highly leveraged, implying a higher required rate of return for debt finance in equilibrium.

Moral-hazard Problems

Moral hazard in R&D investing arises in the usual way: modern industrial firms normally have separation of ownership and management. This leads to a principal–agent problem when the goals of the two conflict, which can result in investment strategies that are not share-value maximizing. Two possible scenarios may coexist: one is the usual tendency of managers to spend on activities that benefit them (growing the firm beyond efficient scale, nicer offices and so on) and the second is a reluctance of risk-averse managers to invest in uncertain R&D projects. Agency costs of the first type may be avoided by reducing the amount of free cash flow available to the managers by leveraging the firm, but this in turn forces them to use the higher-cost external funds to finance R&D (Jensen and Meckling,1976). Empirically, there seem to be limits to the use of the leveraging strategy in R&D-intensive sectors. See Hall (1990, 1994) for evidence that the leveraged buy-out (LBO)/restructuring wave of the 1980s was almost entirely confined to industries and firms where R&D was of no consequence.

According to the second type of principal–agent conflict, managers are more risk-averse than shareholders and avoid R&D projects that will increase the riskiness of the firm. If bankruptcy is a possibility, managers whose opportunity cost is lower than their present earnings and potential bondholders may both wish to avoid variance-increasing projects which shareholders would like to undertake. The argument of the theory is that long-term investments can suffer in this case. The optimal solution to this type of agency cost would be to increase the long-term incentives faced by the manager rather than reducing free cash flow.

Evidence on the importance of agency costs as they relate to R&D takes several forms. Several researchers have studied the impact of anti-takeover amendments (which arguably increase managerial security and willingness to take on risk while reducing managerial discipline) on R&D investment and firm value. Johnson and Rao (1997) find that such amendments are not followed by cuts in R&D, while Pugh *et al.* (1999) find that adoption of an Employee Stock Ownership Plan (ESOP), which is a form of anti-takeover protection, is followed by R&D increases. Cho (1992) finds that R&D intensity increases with the share that managerial shareholdings represent of the manager's wealth, and interprets this as incentive pay mitigating agency costs and inducing long-term investment.

Some have argued that institutional ownership of the managerial firm can reduce the agency costs owing to free-riding by owners that is a feature of the governance of firms with diffuse ownership structure, while others have held that such ownership pays too much attention to short-term earnings and therefore discourages long-term investments. Institutions such as mutual and pension funds often control somewhat larger blocks of shares than individuals, making monitoring firm and manager behaviour a more effective and more rewarding activity for these organizations.

There is some limited evidence that this may indeed be the case. Eng and Shackell (2001) find that firms adopting long-term performance plans for their managers do not increase their R&D spending, but that institutional ownership is associated with higher R&D; R&D firms tend not to be held by banks and insurance companies. Majumdar and Nagarajan (1997) find that high institutional investor ownership does *not* lead to short-term behaviour on the part of the firm; in particular, it does not lead to cuts in R&D spending. Francis and Smith (1995) find that diffusely held firms are less innovative, implying that monitoring alleviates agency costs and enables investment in innovation.

Although the evidence summarized above is fairly clear and indicates that long-term incentives for managers can encourage R&D and that institutional ownership does not necessarily discourage R&D investment, it is fairly silent on the magnitude of these effects, and on whether these governance features truly close the agency cost-induced gap between the cost of capital and the return to R&D. The policy implication, if any, is that nothing in this literature rules out the existence of a wedge between internal and external costs of R&D capital, so there is an argument for some kind of subsidy to R&D investment purely on these grounds, especially for the R&D undertaken by new entrants and firms that do not yet have established cash flow.

Capital Structure and R&D

In the view of some observers, the LBO wave of the 1980s in the USA and the UK arose partly because high real interest rates meant that there were strong pressures to eliminate free cash flow within firms (Blair and Litan, 1990). For firms in industries where R&D is an important form of investment, such pressure should have been reduced by the need for internal funds to undertake such investment, and, indeed, Hall (1993, 1994) and Opler and Titman (1993) find that firms with high R&D intensity were much less likely to do an LBO. Opler and Titman (1994) find that R&D firms that were leveraged suffered more than other firms when facing economic distress, presumably because leverage meant that they were unable to sustain R&D programmes in the face of reduced cash flow.

In related work using data on Israeli firms, Blass and Yosha (2001) report that R&D-intensive firms listed on the US stock exchanges use highly equity-based sources of financing, whereas those listed only in Israel rely more on bank financing and government funding. The former are more profitable and faster growing, which suggests that the choice of where to list the shares and whether to finance with new equity is indeed sensitive to the expected rate of return to the R&D being undertaken. That is, investors supplying arm's-length finance require higher returns to compensate them for the risk of a 'lemon'.

Although leverage may be a useful tool for reducing agency costs in the firm, it is of limited value for R&D-intensive firms. Because the knowledge asset created by R&D investment is intangible, partly embedded in human capital, and ordinarily very specialized to the particular firm in which it resides, the capital structure of R&D-intensive firms customarily exhibits considerably less leverage than those of other firms. Banks and other debtholders prefer to use physical assets to secure loans and are reluctant to lend when the project involves substantial R&D investment rather than investment in plant and equipment. In the words of Williamson (1988), 'redeployable' assets (that is, assets whose value in an alternative use is almost as high as in their current use) are more suited to the governance structures associated with debt. Empirical support for this idea is provided by Alderson and Betker (1996), who find that liquidation costs and R&D are positively related across firms. The implication is that the sunk costs associated with R&D investment are higher than those for ordinary investment.

In addition, servicing debt usually requires a stable source of cash flow, which makes it more difficult to find the funds for an R&D investment programme that must be sustained at a certain level in order to be productive. For both these reasons, firms are either unable or reluctant to use debt finance for R&D investment, which may raise the cost of capital, depending on the precise tax treatment of debt versus equity.[5] Confirming empirical evidence for the idea that limiting free cash flow in R&D firms is a less desirable method of reducing agency costs is provided by Chung and Wright (1998), who find that financial slack and R&D spending are correlated with the value of growth firms positively, but not correlated with those of other firms.

Taxes and the Source of Funds

Tax considerations that yield variations in the cost of capital across source of finance have been well articulated by Auerbach (1984) among others. He argued that, under the US tax system, during most of its history, the cost of financing new investment by debt has been less than that of financing it by retained earnings, which is in turn less than that of issuing new shares. More explicitly, if r is the risk-adjusted required return to capital, τ is the corporate tax rate, θ is the personal tax rate, and c is the capital gains tax rate, we have the following required rates of return for different financing sources:

debt	$r(1-\tau)$	interest deductible at the corporate level;
retained earnings	$r(1-\theta)/(1-c)$	avoids personal tax on dividends, but attracts capital gains tax;
new shares	$r/(1-c)$	eventual capital gains tax.

If dividends are taxed, clearly financing with new shares is more expensive than financing with retained earnings. And unless the personal income tax rate is much higher than the sum of the corporate and capital gains rates, the following inequalities will both hold:

$$(1-\tau)<\frac{1-\theta}{1-c}<\frac{1}{1-c}.$$

These inequalities express the facts that interest expense is deductible at the corporate level, while dividend payments are not, and that shareholders normally pay tax at a higher rate on retained earnings that are paid out than on those retained by the firm and invested.[6] It implicitly assumes that the returns from the investment made will be retained by the firm and eventually taxed at the capital gains rate rather than the rate on ordinary income.

It is also true that the tax treatment of R&D in most OECD economies is very different from that of other kinds of investment: because R&D is expensed as it is incurred, the effective tax rate on R&D assets is lower than that on either plant or equipment, with or without an R&D tax credit in place. This effectively means that the economic depreciation of R&D assets is considerably less than the depreciation allowed for tax purposes – which is 100 per cent – so that the required rate of return for such investment would be lower. In addition, some countries offer a tax credit or subsidy to R&D spending, which can reduce the after-tax cost of capital even further.[7]

The conclusion from this section is that the presence of either asymmetric information or a principal–agent conflict implies that new debt or equity finance will be relatively more expensive for R&D than for ordinary investment, and that considerations such as lack of collateral further reduce the possibility of debt finance. Together, these arguments suggest an important role for retained earnings in the R&D investment decision, independent of their value as a signal of future profitability. In fact, as has been argued by both Hall (1992) and Himmelberg and Petersen (1994), there is good reason to think that positive cash flow may be more important for R&D than for ordinary investment. The next section reports on a series of empirical tests for this proposition.

TESTING FOR FINANCIAL CONSTRAINTS

The usual way to examine the empirical relevance of the arguments that R&D investment in established firms can be disadvantaged when internal funds are not available and recourse to external capital markets required is to estimate R&D investment equations and test for the presence of 'liquidity' constraints, or excess sensitivity to cash-flow shocks. This approach builds on the extensive literature developed for testing ordinary investment equations for liquidity constraints (Fazzari *et al.*, 1988; Arellano and Bond, 1991). It suffers from many of the same difficulties as the estimates in the investment literature, plus one additional problem that arises from the tendency of firms to smooth R&D spending over time.

The ideal experiment for identifying the effects of liquidity constraints on investment is to give firms additional cash exogenously, and observe whether they pass it on to shareholders or use it for investment and/or R&D. If they choose the first alternative, either the cost of capital to the firm has not fallen, or it has fallen but they still have no good investment opportunities. If they choose the second, then the firm must have had some unexploited investment opportunities that were not profitable using more costly external finance. A finding that investment is sensitive to cash-flow shocks that are *not* signals of future demand increases would reject the hypothesis that the cost of external funds is the same as the cost of internal funds. However, lack of true experiments of this kind forces researchers to use econometric techniques such as instrumental variables to attempt to control for demand shocks when estimating the investment demand equation, with varying degrees of success.

The methodology for the identification of R&D investment equations is based on a simple supply and demand heuristic, as shown in Figure 2.1. The curve sloping downward to the right represents the demand for R&D investment funds and the curves sloping upward the supply of funds. Internal funds are available at a constant cost of capital until they are exhausted, at which point it becomes necessary to issue debt or equity in order to finance more investment. When the demand curve cuts the supply curve in the horizontal portion, a shock that increases cash flow (and shifts supply outward) has no effect on the level of investment. However, if the demand curve cuts the supply curve where it is upward sloping, it is possible for a shock to cash flow to shift the supply curve out in such a way as to induce a substantial increase in R&D investment. Figure 2.2 illustrates such a case, where the firm shifts from point A to point B in response to a cash flow shock that does not shift the demand curve.

Econometric work that tests the hypothesis that financing constraints matter for R&D investment has largely been done using standard investment equation methodology. Two main approaches can be identified: one uses a neoclassical accelerator model with ad hoc dynamics to allow for the presence of adjustment costs, and the other an Euler equation derived from

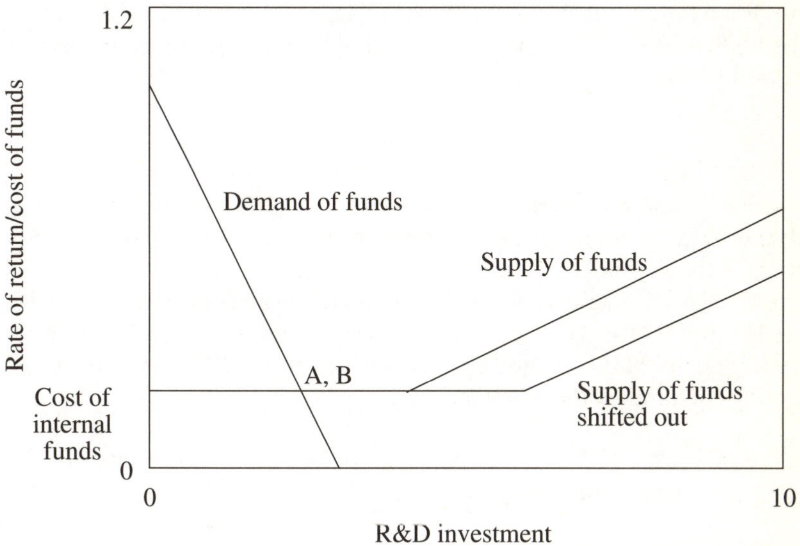

Figure 2.1 Unconstrained firm

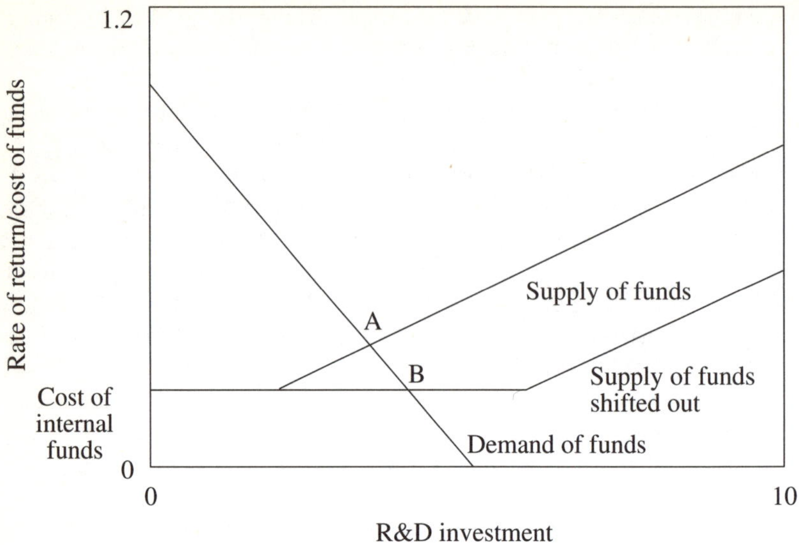

Figure 2.2 Constrained firm

the forward-looking dynamic programme of a profit-maximizing firm that faces adjustment costs for capital.[8]

The accelerator model begins with the marginal product equal to cost condition for capital: $MPK = C$. Assuming that the production function for the ith firm at time t is Cobb–Douglas and taking logarithms of this relationship yields

$$k_{it} = s_{it} + a_i - c_{it}$$

where $k = \log(\text{R\&D capital})$, $s = \log(\text{output or sales})$ and $c = \log(\text{cost of R\&D})$; a_i captures any permanent differences across firms, including differences in the production function.

Lagged adjustment is allowed for by specifying an autoregressive distributed lag (ADL) for the relationship between capital and sales. For example, specifying an ADL(2,2) and approximating Δk by $R/K - \delta$ yields an estimating equation of the following form:

$$R/K = f(R(-1)/K(-1), \Delta s, \Delta s(-1), k(-2) - s(-2), \text{time dummies, firm dummies}).$$

The time dummies capture the conventional cost of capital, assumed to be the same for all firms. Firm-specific costs related to financing constraints are included by adding current and lagged values of the cash flow/capital ratio to this equation. Because of the presence of firm dummies, estimation is done using first differences of this equation, instrumented by lagged values of the right-hand-side variables to correct for the potential endogeneity of the contemporaneous values. In principle, this will also control for the potential simultaneity between current investment and the disturbance.

The Euler-equation approach begins with the following first-order condition for investment in two adjacent periods:

$$E_{t-1}, [MPK_t + (1-\delta)(p_t + MAC_t) - (1+r)(\alpha_{t-1}/\alpha_t)\,(p_{t-1} + MAC_{t-1}] = 0,$$

where MAC denotes the marginal adjustment costs for capital and α_t is the shadow value of investment funds in period t, which will be unity if there are no financing constraints. After specifying a Cobb–Douglas production function and quadratic adjustment costs, we obtain the following estimating equation:

$$R/K = f(R(-1)/K(-1),\ S/K,\ (R/K)^2,\ \text{time dummies, firm dummies}).$$

Like the accelerator model, this equation should also be estimated in differenced form with lagged values of the right-hand-side variables as instruments.

When financial constraints are present, the coefficient of lagged R&D investment in the Euler equation differs from $(1+r)$ by the term (α_{t-1}/α_t). The implication is that, when the firm changes its financial position (that is, the shadow value of additional funds for investment changes) between one period and the next, it will invest as though it is facing a cost of capital greater than r (when the shadow value falls between periods) or less than r (when the shadow value rises between periods). Clearly this is a very difficult test to perform because (α_{t-1}/α_t) is not constant across firms or across time periods, so it cannot be treated as a parameter.

Three solutions are possible: the first is to model (α_{t-1}/α_t) as a function of proxies for changes in financial position, such as dividend behaviour, new share issues, or new debt issues. The second is more ad hoc: recall that this term also multiplies the price p_t, of R&D capital to create a firm-specific cost of capital. Most researchers simply include the cash-flow/capital ratio in the model to proxy for the firm-specific cost of capital and test whether it enters in the presence of time dummies that are the same for all firms. This method assumes that all firms face the same R&D price (cost of capital), except for the cash-flow effect.

The third possibility is to stratify firms in some way that is related to the level of cash constraints that they face (for example, dividend-paying and non-dividend-paying firms), estimate separate investment equations for each group, and test whether the coefficients are equal. This last was the method used by Fazzari *et al.* (1988) in the paper that originated this literature. Note that this approach does not rely on the full Euler-equation derivation given above, but uses a version of the neoclassical accelerator model (the first model given above).

During the past few years, various versions of the methodologies described above have been applied to data on the R&D investment of US, UK, French, German, Irish and Japanese firms. The firms examined are typically the largest and most important manufacturing firms in their economy. For example, Hall (1992) found a large positive elasticity between R&D and cash flow, using an accelerator-type model and a very large sample of US manufacturing firms. The estimation methodology here controlled for both firm effects and simultaneity. Similarly, and using some of the same data, Himmelberg and Petersen (1994) looked at a panel of 179 US small firms in high-tech industries and found an economically large and statistically significant relationship between R&D investment and internal finance.

Harhoff (1998) found weak but significant cash-flow effects on R&D for both small and large German firms, although Euler-equation estimates for R&D investment were uninformative owing to the smoothness of R&D and the small sample size. Combining limited survey evidence with his regression results, he concludes that R&D investment in small German firms may be constrained by the availability of finance. Bond *et al.* (1999) find significant differences between the cash-flow impacts on R&D and investment for large manufacturing firms in the UK and Germany. German firms in their sample are insensitive to cash-flow shocks, whereas the investment of UK firms which do not engage in R&D does respond. Cash flow helps to predict whether a UK firm carries out R&D, but not the level of that R&D. They interpret their findings to mean that financial constraints are important for British firms, but that those which carry out R&D are a self-selected group that face fewer constraints. This is consistent with the view that the desire of firms to smooth R&D over time combines with the relatively high cost of financing it to reduce R&D well below the level that would obtain in a frictionless world.

Mulkay *et al.* (2001) perform a similar exercise using large French and US manufacturing firms, finding that cash-flow impacts are much larger in the USA than in France, both for R&D and for ordinary investment. Except for the well-known fact that R&D exhibits higher serial correlation than investment (presumably because of higher adjustment costs), differences in behaviour are between countries, not between investment types. This result is consistent with evidence reported in Hall *et al.* (1999) for the USA, France and Japan during an earlier time period, which basically finds that R&D and investment on the one hand, and sales and cash flow on the other, are simultaneously determined in the USA (neither one 'Granger-causes' the other), whereas in the other countries, there is little feedback from sales and cash flows to the two investments. Using a non-structural R&D investment equation together with data for the USA, UK, Canada, Europe and Japan, Bhagat and Welch (1995) found similar results for the 1985–90 period, with stock returns predicting changes in R&D more strongly for the US and UK firms.

Recently, Bougheas *et al.* (2001) examined the effects of liquidity constraints on R&D investment using firm-level data for manufacturing firms in Ireland and also found evidence that R&D investment in these firms is financially constrained, in line with the previous studies of US and UK firms.

Brown (1997) argues that existing tests of the impact of capital-market imperfections on innovative firms cannot distinguish between two possibilities: (i) capital markets are perfect and different factors drive the firm's different types of expenditure, or (ii) capital markets are imperfect and different types of expenditure react differently to a common factor (shocks to the supply of internal finance). He then compares the sensitivity of investment to cash flow for innovative and non-innovative firms. The results support the hypothesis that capital markets are imperfect, finding that the investment of innovative firms is more sensitive to cash flow.

The conclusions from this body of empirical work are several: first, there is solid evidence that debt is a disfavoured source of finance for R&D investment; second, the 'Anglo-Saxon' economies, with their thick and highly developed stock markets and relatively transparent ownership structures, typically exhibit more sensitivity and responsiveness of R&D to cash flow than Continental economies; third, and much more speculatively, this greater responsiveness may arise because they are financially constrained, in the sense that they view external sources of finance as much more costly than internal, and therefore require a considerably higher rate of return to investments done on the margin when they are tapping these sources. However, it is perhaps equally likely that this responsiveness occurs because firms are more sensitive to demand signals in thick financial equity markets; a definitive explanation of the 'excess sensitivity' result awaits further research.[9] In addition to these results, the evidence from Germany and some other countries suggests that small firms are more likely to face this difficulty than large established firms (not surprisingly, if the source of the problem is a 'lemons' premium).

From a policy perspective, these results point to another reason why it may be socially beneficial to offer tax incentives to companies in order to reduce the cost of capital they face for R&D investment, especially to small and new firms. Many governments, including those in the USA and the

UK, currently have such programmes. Such a policy approach simply observes that the cost of capital is relatively high for R&D and tries to close the gap via a tax subsidy. However, there is an alternative approach relying on the private sector that attempts to close the financing gap by reducing the degree of asymmetric information and moral hazard rather than simply subsidizing the investment. I turn to this topic in the next section.

SMALL FIRMS, START-UP FINANCE AND VENTURE CAPITAL

As should be apparent from much of the preceding discussion, any problems associated with financing investments in new technology will be most apparent for new entrants and start-up firms. For this reason, many governments already provide some form of assistance for such firms, and in many countries, especially the USA, there exists a private sector 'venture capital' (VC) industry that is focused on solving the problem of financing innovation for new and young firms. This section reviews what we know about these alternative funding mechanisms, beginning with a brief look at government funding for start-ups and then discussing the venture capital solution.

Government Funding for Start-up Firms

Examples of such programmes are the US Small Business Investment Company (SBIC) and Small Business Innovation Research (SBIR) programmes. Together, these programmes disbursed $2.4 billion in 1995, more than 60 per cent of the amount from VC in that year (Lerner, 1998a). In Germany, more than 800 federal and state government financing programmes have been established for new firms in the recent past (OECD, 1995). In 1980, the Swedish established the first of a series of investment companies (along with instituting a series of measures such as reduced capital-gains taxes to encourage private investments in start-ups), partly on the US model. By 1987, the government share of venture capital funding was 43 per cent (Karaomerliolu and Jacobsson,1999). Recently, the UK has instituted a series of government programmes under the Enterprise Fund umbrella, which allocate funds to small and medium-sized firms in high-technology and certain regions, as well as guaranteeing some loans to small businesses (Bank of England, 2001). There are also programmes at the European level.

A limited amount of evidence, most of it USA-based, exists as to the effectiveness and 'additionality' of these programmes. In most cases, evaluating the success of the programmes is difficult owing to the lack of a 'control' group of similar firms that do not receive funding.[10] Therefore, most of the available studies are based on retrospective survey data provided by the recipients; few attempt to address seriously the question of performance under the counterfactual. A notable exception is the study by Lerner (1999), who looks at 1435 SBIR awardees and a matched sample of firms that did not receive awards over a 10-year post-award period. Because most of the firms are privately held, he is unable to analyse the resulting valuation or profitability of the firms, but he does find that firms receiving SBIR grants grow significantly faster than the others after receipt of the grant. He attributes some of this effect to 'quality certification' by the government that enables the firm to raise funds from private sources as well.[11]

Venture Capital

Many observers view the rise of the VC industry, especially that in the USA, as a 'free market' solution to the problems of financing innovation. In fact, many of the European programmes described above have as some of their goals the provision of seed capital and the encouragement of a VC industry that addresses the needs of high-technology start-ups. Table 2.1 shows why this has been a policy concern: the amount of VC available to firms in the USA and Europe was roughly comparable in 1996, but the relative allocation to new firms (seed money and start-ups) in Europe was much less, below 10 per cent of the funds as opposed to 27 per cent. A correspondingly greater amount was used to finance buy-outs of various kinds.

Table 2.1 Venture capital disbursements by stage of financing, 1996

	USA	Europe
Total VC disbursements ($m. 1996)	9420.6	8572.0
Share, seed and start-ups (%)	27.1	6.5
Share for expansion (%)	41.6	39.3
Share, other (including buy-outs) (%)	31.3	54.2

Source: Rausch (1998) and author's calculations.

In the USA, the VC industry consists of fairly specialized pools of funds (usually from private investors) that are managed and invested in companies by individuals knowledgeable about the industry in which they are investing. In principle, the idea is that the lemons premium is reduced because the investment managers are better informed, and moral hazard is minimized because a higher level of monitoring than that used in conventional arm's-length investments is the norm. But the story is more complex than that: the combination of high uncertainty, asymmetric information, and the fact that R&D investment typically does not yield results instantaneously, not only implies option-like behaviour for the investment decision, but also has implications for the form of the VC contract and the choice of decision maker. That is, there are situations in which it is optimal for the investor (the venture capitalist) to have the right to shut down a project and there are other situations in which optimal performance is achieved when the innovator has control.

A number of studies have documented the characteristics and performance of the VC industry in the USA. The most detailed look at the actual operation of the industry is that by Kaplan and Stromberg (2000), who examine 200 VC contracts and compare their provisions to the predictions of the economic theory of financial contracting under uncertainty. They find that the contracts often provide for separate allocation of cash-flow rights, control rights, voting rights, board positions and liquidation rights, and that the rights are frequently contingent on performance measures. If performance is poor, the venture capitalists often gain full control of the firm. Provisions such as delayed vesting are often included to mitigate hold-up by the entrepreneur as suggested by Anand and Galetovic (2000).

Kaplan and Stromberg conclude that these contracts are most consistent with the predictions of Aghion and Bolton (1992) and Dewatripont and Tirole (1994), all of whom study the incomplete contracts that arise when cash flows can be observed but not verified in sufficient detail to be used for contract enforcement. Put simply, the modal VC contract is a complex debt–equity hybrid (and, in fact, frequently contains convertible preferred securities and other such instruments) that looks more like debt when the firm does poorly (giving control to the investor) and more like equity when the firm does well (by handing control to the entrepreneur, which is incentive-compatible).

In a series of papers, Lerner (1992, 1995) studied a sample of VC-financed start-ups in detail, highlighting the important role that investing and monitoring experience has in this industry. He found that the amount of funds provided and the share of equity retained by the managers are sensitive to the experience and ability of the capital providers and the maturity of the firm being funded. Venture capitalists do increase the value of the firms they fund, especially when they are experienced investors. Firms backed by seasoned VC financiers are more likely successfully to time the market when they go public, and to employ the most reputable underwriters.

At a macroeconomic level, VC funding tends to be pro-cyclical, but it is difficult to disentangle whether the supply of funding causes growth, or productivity growth encourages funding (Gompers and Lerner, 1999a,b;

Kortum and Lerner, 2000; Ueda, 2001). The problem here is very similar to the identification problem for R&D investment in general: because of feedback effects, there is a chicken–egg simultaneity in the relationship. Some evidence (Majewski, 1997) exists that new and/or small biotechnology firms turn to other sources of funding in downturns, but that such placements are typically less successful (Lerner and Tsai, 2000), owing to the misallocation of control rights (when the start-up firm is in a weak bargaining position, control tends to be allocated to the more powerful corporate partner, but this has negative consequences for incentives).

The limited evidence from Europe on the performance of VC-funded firms tends to confirm that from the USA. Engel (2001) compares a matched sample of German firms founded between 1991 and 1998 and finds that the VC-backed firms grew faster than the non-VC-backed firms. Lumme *et al.* (1993) compare the financing and growth of small UK and Finnish firms. This approach permits a comparison between a financial market-based and a bank-centred economy and, indeed, they find that small UK firms rely more on equity and less on loan finance and grow faster than small Finnish firms. Further evidence on small UK high-technology firms is provided by Moore (1993), who looks at 300 such firms, finding that the availability and cost of finance is the most important constraint facing these firms, but that they are affected only marginally more than other types of small firms. That is, the financing 'gap' in the UK may be more related to size than to R&D intensity.

For Japan, Hamao *et al.* (1998) find that the long-run performance of VC-backed initial public offerings (IPOs) is no better than that of other IPOs, unlike Lerner's evidence for the USA. However, many VCs in Japan are subsidiaries of major securities firms rather than specialists, as in the USA. Only these VCs have low returns, whereas those that are independent have returns more similar to the US. They attribute the low returns to conflicts of interest between the VC subsidiary and the securities firm that owns it, which affects the price at which the IPO is offered. This result highlights the importance of the institutions in which the VC industry is embedded for the creation of entrepreneurial incentives.

Black and Gilson (1998) and Rajan and Zingales (2001) take the institutional argument further. Both pairs of authors emphasize the contrast between arm's-length market-based financial systems (such as the USA and the UK) and bank-centred capital market systems (such as much of Continental Europe and Japan), and view VC as combining the strengths of the two systems, in that it provides both the strong incentives for the manager–entrepreneur characteristic of the stock-market system and the monitoring by an informed investor characteristic of the bank-centred system. They emphasize the importance of an active stock market, especially for newer and younger firms, in order to provide an exit strategy for VC investors, and allow them to move on to financing new start-ups. Thus having a VC industry that contributes to innovation and growth requires the existence of an active IPO market to permit successful entrepreneurs to regain control of their firms (and, incidentally, to provide powerful incentives for undertaking the start-up in the first place) and also to ensure that the VCs themselves are able to use their expertise to help to establish new endeavours.

CONCLUSIONS

On the literature surveyed here, what do we know about the necessity and effectiveness of possible policy options towards the financing of R&D? Several main points emerge.

- There is fairly clear evidence, based on theory, surveys and empirical estimation, that small and start-up firms in R&D-intensive industries face a higher cost of capital than their larger competitors and than firms in other industries. In addition to compelling theoretical arguments and empirical evidence, the mere existence of the VC industry and the fact that it is concentrated precisely where these start-ups are most active suggests that this is so. In spite of considerable entry into the VC industry, returns remain high, which does suggest a high required rate of return in equilibrium (*Upside*, 2001).
- The evidence for a financing gap for large and established R&D firms is harder to establish. It is certainly the case that these firms prefer to use internally generated funds for financing investment, but less clear that there is an important role for policy, beyond the favourable tax treatment that currently exists in many countries.[12]
- The VC solution to the problem of financing innovation has its limits. First, it does tend to focus only on a few sectors at a time, and to make investment with a minimum size that is too large for start-ups in some fields. Second, good performance of the VC sector requires a thick market in small and new firm stocks (such as NASDAQ or EASDAQ) in order to provide an exit strategy for early-stage investors.
- The effectiveness of government incubators, seed funding, loan guarantees and other such policies for funding R&D deserves further study, ideally in an experimental or quasi-experimental setting. In particular, studying the cross-country variation in the performance of such programmes would be desirable, because the outcomes may depend to a great extent on institutional factors that are difficult to control for using data from within a single country.

NOTES

1. I am grateful to Colin Mayer and Andrew Glyn for very helpful comments on the first draft.
2. See, for example, Schumpeter (1942, ch. 8, fn. 1).
3. See Hall and Van Reenen (2000) for details. The USA has an incremental R&D tax credit with a value for A^c of about 0.13, whereas the UK has no credit at the present time, so $A^c = 0$.
4. Since 1974, publicly traded firms in the USA have been required to report their total R&D expenditures in their annual reports and 10-K filings with the Securities and Exchange Commission, under Financial Accounting Standards Board rule no. 2, issued in October 1974. In 1989, a new accounting standard, SSAP 13, made similar disclosures obligatory in the UK. Most Continental European countries do not have such a requirement, although they may evolve in that direction owing to international harmonization of accounting standards, at least for publicly traded firms.
5. There is also considerable cross-sectional evidence for the USA that R&D intensity and leverage are negatively correlated across firms – see Friend and Lang (1988), Hall (1992) and Bhagat and Welch (1995).
6. A detailed discussion of tax regimes in different countries is beyond the scope of this survey, but it is quite common in several countries for long-term capital gains on funds that remain with a firm for more than one year to be taxed at a lower rate than ordinary income. Of course, even if the tax rates on the two kinds of income are equal, the inequalities will hold. Only in the case where dividends are not taxed at the corporate level (which was formerly the case in the UK) will the ranking given above not hold.
7. See Hall and Van Reenen (2000) for details.
8. A detailed consideration of the econometric estimation of these models can be found in Mairesse *et al.* (1999). See also Hall (1991).
9. It is also true that much of the literature here has tended to play down the role of measurement error in drawing conclusions from the results. Measurement error in Tobin's q, cash flow, or output is likely to be sizeable and will ensure that all variables will enter any specification of the R&D investment equation significantly, regardless of whether they truly belong or not. Instrumental variables estimation is a partial solution, but only if all the errors are serially uncorrelated, which is unlikely.
10. See Jaffe (2002) for a review of methodologies for evaluation of such government programmes. For a complete review of the SBIR programme, including some case studies, see the National Research Council (1998).
11. Also see Spivack (2001) for further studies of such programmes, including European studies, and David *et al.* (2000) and Klette *et al.* (2000) for surveys of the evaluation of government R&D programmes in general.
12. It is important to remind the reader of the premise of this chapter: I am focusing *only* on the financing-gap arguments for favourable treatment of R&D and ignoring (for the present) the arguments based on R&D spillovers and externalities. There is good reason to believe that the latter are a much more important consideration for large established firms, especially if we wish those firms to undertake basic research that is close to industry but with unknown applications (the Bell Labs model).

REFERENCES

Aghion, P. and P. Bolton (1992), 'An Incomplete Contracts Approach to Financial Contracting', *Review of Economic Studies*, **77**, 338–401.

Aghion, P. and P. Howitt (1997), *Endogenous Growth Theory*, Cambridge, MA: MIT Press.

Akerlof, G.A. (1970), 'The Market for "Lemons": Quality, Uncertainty, and the Market Mechanism', *Quarterly Journal of Economics*, **84**, 488–500.

Alam, P. and K.S. Walton (1995), 'Information Asymmetry and Valuation Effects of Debt Financing', *Financial Review*, **30**(2), 289–311.

Alderson, M.J. and B.L. Betker (1996), 'Liquidation Costs and Accounting Data', *Financial Management*, **25**(2), 25–36.

Anand, B.N. and A. Galetovic (2000), 'Weak Property Rights and Holdup in R&D', *Journal of Economics and Management Strategy*, **9**(4), 615–42.

Anton, J.J. and D.A. Yao (1998), 'The Sale of Intellectual Property: Strategic Disclosure, Property Rights, and Incomplete Contracts', working paper, Wharton School, University of Pennsylvania.

Arellano, M. and S. Bond (1991), 'Some Tests of Specification for Panel Data: Monte Carlo Evidence and an Application to Employment Equations', *Review of Economic Studies*, **58**, 277–97.

Arrow, K.J. (1962), 'Economic Welfare and the Allocation of Resources for Invention', in R. Nelson (ed.), *The Rate and Direction of Inventive Activity*, Princeton, NJ: Princeton University Press.

Auerbach, A.J. (1984), 'Taxes, Firm Financial Policy, and The Cost of Capital: An Empirical Analysis', *Journal of Public Economics*, **23**, 27–57.

Bank of England (2001), *Finance for Small Firms – An Eighth Report*, London: Domestic Finance Division, Bank of England.

Bhagat, S. and I. Welch (1995), 'Corporate Research and Development Investments: International Comparisons', *Journal of Accounting and Economics*, **19**(March–May).

Bhattacharya, S. and J.R. Ritter (1983), 'Innovation and Communication: Signaling with Partial Disclosure', *Review of Economic Studies*, **50**, 331–46.

Black, B.S. and R.J. Gilson (1998), 'Venture Capital and the Structure of Capital Markets: Banks versus Stock Markets', *Journal of Financial Economics*, **47**(3), 243–77.

Blair, M.M. and R.E. Litan (1990), *Corporate Leverage and Leveraged Buyouts in the Eighties*, Washington, DC: Brookings Institution.

Blass, A.A. and O. Yosha (2001), 'Financing R&D in Mature Companies: An Empirical Analysis', working paper, Bank of Israel, Tel Aviv University and CEPR.

Bond, S., D. Harhoff and J. Van Reenen (1999), 'Investment, R&D, and Financial Constraints in Britain and Germany', working paper no. 99/5, Institute of Fiscal Studies, London.

Bougheas, S., H. Goerg and E. Strobl (2001), 'Is R&D Financially Constrained? Theory and Evidence from Irish Manufacturing', University of Nottingham.

Brown, W. (1997), 'R&D Intensity and Finance: Are Innovative Firms Financially Constrained?', London School of Economics Financial Market Group.

Cho, S. (1992), 'Agency Costs, Management Stockholding, and Research and Development Expenditures', *Seoul Journal of Economics*, **5**(2), 127–52.

Chung, K.H. and P. Wright (1998), 'Corporate Policy and Market Value: A q Theory Approach', *Review of Quantitative Finance and Accounting*, **11**(3), 293–310.

David, P.A., B.H. Hall and A.A. Toole (2000), 'Is Public R&D a Complement or a Substitute for Private R&D? A Review of the Econometric Evidence', *Research Policy*, **29**, 497–530.

Dewatripont, M. and J. Tirole (1994), 'A Theory of Debt and Equity: Diversity of Securities and Manager–Shareholder Congruence', *Quarterly Journal of Economics*, **109**, 1027–54.

Eng, L.L. and M. Shackell (2001), 'The Implications of Long-term Performance

Plans and Institutional Ownership for Firms' Research and Development Investments', *Journal of Accounting, Auditing and Finance*, **16**(2), 117–39.
Engel, D. (2001), 'Hoeheres Beschaeftigungswachstum Durch Venture Capital?', ZEW Discussion Paper No. 01–34, Mannheim.
Fazzari, S.M., R.G. Hubbard and B.C. Petersen (1988), 'Financing Constraints and Corporate Investment', *Brookings Papers on Economic Activity*, **1**, 141–205.
Francis, J. and A. Smith (1995), 'Agency Costs and Innovation: Some Empirical Evidence', *Journal of Accounting and Economics*, **19**(2/3), 383–409.
Friend, I. and H.H.P. Lang (1988), 'An Empirical Test of the Impact of Management Self-interest on Corporate Capital Structure', *Journal of Finance*, **43**, 271–83.
Gompers, P.A. and J. Lerner (I999a), 'What Drives Venture Capital Fundraising?', NBER Working Paper No. 6906, Cambridge, MA.
Gompers, P.A. (1999b), *Capital Formation and Investment in Venture Markets: Implications for the Advanced Technology Program*, Washington, DC: Advanced Technology Program, NIST, US Department of Commerce.
Griliches, Z. (1992), 'The Search for R&D Spillovers', *Scandinavian Journal of Economics*, **94**, S29–47.
Hall, B.H. (1990), 'The Impact of Corporate Restructuring on Industrial Research and Development', *Brookings Papers on Economic Activity*, **1**, 85–136.
Hall, B.H. (1991), 'Firm-level Investment with Liquidity Constraints: What Can the Euler Equations Tell Us?', mimeo, University of California at Berkeley and the National Bureau of Economic Research.
Hall, B.H. (1992), 'Research and Development at the Firm Level: Does the Source of Financing Matter?', NBER Working Paper no. 4096, June.
Hall, B.H. (1993), 'R&D Tax Policy During the Eighties: Success or Failure?', *Tax Policy and the Economy*, **7**, 1–36.
Hall, B.H. (1994), 'Corporate Capital Structure and Investment Horizons in the United States, 1976–1987', *Business History Review*, **68**, 110–43.
Hall, B.H. (1996), 'The Private and Social Returns to Research and Development', in B.L.R. Smith and C.E. Barfield (eds), *Technology, R&D, and the Economy*, Washington, DC: Brookings Institution and the American Enterprise Institute, pp. 140–83.
Hall, B.H. and J. Van Reenen (2000), 'How Effective are Fiscal Incentives for R&D? A New Review of the Evidence', *Research Policy*, **29**, 449–69.
Hall, B.H., Z. Griliches and J.A. Hausman (1986), 'Patents and R&D: Is There a Lag?', *International EconomicReview*, **27**, 265–83.
Hall, B.H., J. Mairesse, L. Branstetter and B. Crepon (1999), 'Does Cash Flow Cause Investment and R&D? An Exploration using Panel Data for French, Japanese, and United States Firms in the Scientific Sector', in D. Audretsch and A.R. Thurik (eds), *Innovation, Industry Evolution and Employment*, Cambridge: Cambridge University Press.
Hamao, Y., F., Packer and J.R. Ritter (1998), 'Institutional Affiliation and the Role of Venture Capital: Evidence from Initial Public Offerings in Japan', Federal Reserve Bank of New York Staff Report no. 52.
Harhoff, D. (1998), 'Are There Financing Constraints for R&D and Investment in German Manufacturing Firms?', *Annales d Economie et de Statistique*, **49/50**, 421–56.
Himmelberg, C.P. and B.C. Petersen (1994), 'R&D and Internal Finance: A Panel Study of Small Firms in High-tech Industries', *Review of Economics and Statistics*, **76**, 38–51.

Jaffe, A.B. (2002), 'Building Programme Evaluation into the Design of Public Research-Support Programmes', *Oxford Review of Economic Policy*, **18**(1), 22–34.

Jensen, M.C. and W. Meckling (1976), 'Theory of the Firm: Managerial Behavior, Agency Costs, and Ownership Structure', *Journal of Financial Economics*, **3**, 305–60.

Johnson, M.S. and R.P. Rao (1997), 'The Impact of Antitakeover Amendments on Corporate Financial Performance', *Financial Review*, **32**(4), 659–89.

Kaplan, S.N. and P. Stromberg (2000), 'Financial Contracting Theory Meets the Real World: An Empirical Analysis of Venture Capital Contracts', NBER Working Paper no. 7660, Cambridge, MA.

Karaomerliolu, D.C. and S. Jacobsson (1999), 'The Swedish Venture Capital Industry – An Infant, Adolescent, or Grown-up?', Chalmers Institute of Technology, Gottenburg, Sweden.

Klette, T.J., J. Møen and Z. Griliches (2000), 'Do Subsidies to Commercial R&D Reduce Market Failures? Microeconometric Evaluation Studies', *Research Policy*, **29**, 471–96.

Kortum, S. and J. Lerner (2000), 'Assessing the Contribution of Venture Capital to Innovation', *RAND Journal of Economics*, **31**(4), 674–92.

Lach, S. and M. Schankerman (1988), 'Dynamics of R&D and Investment in the Scientific Sector', *Journal of Political Economy*, **97**(4), 880–904.

Leland, H.E. and D.H. Pyle (1977), 'Informational Asymmetries, Financial Structure, and Financial Intermediation', *Journal of Finance*, **32**, 371–87.

Lerner, J. (1992), 'Venture Capitalists and the Decision to Go Public', Working Paper no. 93-002, Harvard Business School.

Lerner, J. (1995), 'Venture Capitalists and the Oversight of Privately Held Firms', *Journal of Finance*, **50**, 301–18.

Lerner, J. (1998a), '"Angel" Financing and Public Policy: An Overview', *Journal of Banking and Finance*, **22**, 773–83.

Lerner, J. (1998b), '"Public Venture Capital": Rationale and Evaluation', in National Research Council (ed.), *SBIR: Challenges and Opportunities*, Washington, DC: Board of Science, Technology and Economic Policy, NRC.

Lerner, J. (1999), 'The Government as Venture Capitalist: The Long-run Effects of the SBIR Program', *Journal of Business*, **72**, 285–318.

Lerner J. and A. Tsai (2000), 'Do Equity Financing Cycles Matter? Evidence from Biotechnology Alliances', NBER Working Paper no. 7464, January.

Levin, R.C., A.K. Klevorick, R.R. Nelson and S.G. Winter (1987), 'Appropriating the Returns from Industrial Research and Development', *Brookings Papers on Economic Activity*, **3**, 783–832.

Lumme, A. *et al.* (1993), 'New, Technology-based Companies in Cambridge in an International Perspective', University of Cambridge Small Business Research Centre Working Paper 35, September.

Mairesse, J., B.H. Hall and B. Mulkay (1999), 'Firm-level Investment in France and the United States: An Exploration of What we have Learned in Twenty Years', *Annales d'Economie et de Statistique*, **55–6**, 27–69.

Majewski, S.E. (1997), 'Using Strategic Alliance Formation as a Financing Mechanism in the Biotechnology Industry', mimeo, UCBerkeley.

Majumdar, S.K. and A. Nagarajan (1997), 'The Impact of Changing Stock Ownership Patterns in the United States: Theoretical Implications and Some Evidence', *Revue d'Economie Industrielle*, **82**, 39–54.

Mansfield, E., M. Schwartz and S. Wagner (1981), 'Imitation Costs and Patents: An Empirical Study', *The Economic Journal*, **91**, 907–18.
Miller, M.H. and F. Modigliani (1961), 'Dividend Policy, Growth, and the Valuation of Shares', *Journal of Business*, **34**, 411–33.
Modigliani, F. and M.H. Miller (1958), 'The Cost of Capital, Corporation Finance and the Theory of Investment', *American Economic Review*, **48**, 261–97.
Moore, B. (1993), 'Financial Constraints to the Growth and Development of Small High-technology Firms', University of Cambridge Small Business Research Centre Working Paper 31, July.
Mulkay, B., B.H. Hall and J. Mairesse (2001), 'Investment and R&D in France and in the United States', in Deutsche Bundesbank (ed.), *Investing Today for the World of Tomorrow*, New York: Springer.
National Research Council (1998), *SBIR: Challenges and Opportunities*, Washington, DC: Board on Science, Technology and Economic Policy, NRC.
Nelson, R.R. (1959), 'The Simple Economics of Basic Scientific Research', *Journal of Political Economy*, **49**, 297–306.
OECD (1995), *Venture Capital in OECD Countries*, Paris: Organization for Economic Cooperation and Development.
Opler, T.C. and S. Titman (1993), 'The Determinants of Leveraged Buyout Activity: Free Cash Flow vs. Financial Distress Costs', *Journal of Finance*, **48**(5), 1985–9.
Opler, T.C. (1994),' Financial Distress and Corporate Performance', *Journal of Finance*, **49**(3), 1015–10.
Pugh, W.N., J.S. Jahera Jr and S. Oswald (1999), 'ESOPs, Takeover Protection, and Corporate Decision Making', *Journal of Economics and Finance*, **23**(2), 170–83.
Rajan, R.G. and L. Zingales (2001), 'Financial Systems, Industrial Structure, and Growth', *Oxford Review of Economic Policy*, **17**(4), 467–82.
Rausch, L.M. (1998), 'Venture Capital Investment Trends in the United States and Europe', National Science Foundation Division of Science Resource Studies Issues Brief 99–303, Washington, DC.
Romer, P.M. (1986), 'Increasing Returns and Long Run Growth', *Journal of Political Economy*, **94**(5), 1002–37.
Scherer, F.M. (1998), 'The Size Distribution of Profits from Innovation', *Annales d Economie et de Statistique*, **49/50**,495–516.
Schumpeter, J. (1942), *Capitalism, Socialism, and Democracy*, New York: Harper & Row (reprinted 1960).
Spivack, R.N. (2001), *The Economic Evaluation of Technological Change*, Conference Proceedings ofthe Advanced Technology Program, Washington, DC, National Institute of Standards and Technology.
Szewczyk, S.H., G.P. Tsetsekos and Z.Z. Zantout (1996), 'The Valuation of Corporate R&D Expenditures: Evidence from Investment Opportunities and Free Cash Flow', *Financial Management*, **25**(1), 105–10.
Ueda, M. (2001), 'Does Innovation Spur Venture Capital?', Barcelona, Universitat Pompeu Fabra.
Upside (2001), December.
Williamson, O.E. (1988), 'Corporate Finance and Corporate Governance', *Journal of Finance*, **43**, 567–91.
Zantout, Z.Z. (1997), 'A Test of the Debt Monitoring Hypothesis: The Case of Corporate R&D Expenditures', *Financial Review*, **32**(l), 21–48.

3. The financing and governance of new technologies

Colin Mayer*

1. INTRODUCTION

What are the financial sector preconditions for the successful development of a high-technology sector? The conventional answer is straightforward: an active venture capital industry combined with a liquid domestic stock market. The development of venture capital firms and stock markets is regarded as a priority for the growth of high-tech industries. Is this correct? How do high-tech firms finance themselves and what role do stock markets play in their development?

There is accumulating evidence of a relationship between financial development and economic growth. Several studies report a relation between the size of financial systems at the start of a period and subsequent economic growth. Controlling for other considerations, financial development appears to contribute to growth. A range of measures of financial development are relevant – the volume of monetary assets, the size of banking systems and the size of stock markets.

To the extent that it is possible to establish the channel by which financial development contributes to growth, it appears to be through the external financing of firms. Comparing the growth of different industries across countries or different companies suggests that there is an interrelationship between their growth rates, the extent to which they are dependent on external finance and the development of financial systems in which they are operating. In other words, financial development confers particular advantages on industries and companies that are especially dependent on external finance.

These results are consistent with the view that a primary function of financial institutions is to improve allocation of funds within an economy.

* Reprinted from *Information Economics and Policy*, **14**, Mayer, C., 'Financing the New Economy: Financial institutions and corporate governance', pp. 311–326, 2002 with permission from Elsevier.

Corporate, industrial and economic growth are assisted by institutions that direct financing to activities that are most dependent on external finance. The studies therefore provide empirical confirmation at an aggregate or industry level of the theoretical underpinning of financial institutions.

However, the question that these studies leave unanswered is which institutions are particularly well suited to performing these functions. Do all institutions serve companies equally well or are some institutions particularly well-suited to the financing of high technology?

The second set of issues concerns the policies that can be used to influence the development of institutions. Over the last few years a literature has emerged emphasizing the important role that legal and regulatory structures play in influencing institutional development. This literature has emphasized protection of investors as being a crucial determinant of the development of financial systems. Since, as noted above, the development of financial systems is in turn related to the external financing of firms, this points to a key role for investor protection in promoting the external financing and growth of firms. The policy message that appears to emerge from these studies is clear: improve investor, in particular minority investor, protection, and financial development, investment and growth will follow.

This raises the question of what precisely is the relation between legal systems, regulation and the structure of financial institutions. Is there, as the above literature suggests, a straightforward relation between regulation and the development of institutions? In particular, are certain regulatory rules suited to the financing of high-technology activities?

Section 2 of this chapter reviews evidence on comparative financial systems. Section 3 discusses ownership and control. Section 4 describes emerging theories that point to a comparative advantage of different financial and governance systems in promoting particular types of activities. The chapter then turns to an illustration of this in the context of high-technology industries. Section 5 discusses first, the pre-initial public offering (IPO) stage, and then post-IPOs. Section 6 considers the relation between financial institutions and venture capital financing and in particular how sources of venture capital funding are related to their activities. Section 7 considers the role of the state. Section 8 discusses policy implications and section 9 concludes the chapter.

2. COMPARATIVE FINANCIAL SYSTEMS

There has been extensive comparison of the performance of different financial systems.[1] These analyses have focused on the contrast between bank-oriented and market-oriented systems. Most of the studies compare a small number of countries, focusing in particular on the UK and USA on the one hand, and Germany and Japan on the other.[2]

The criteria by which systems are categorized include corporate financing, bank ownership of corporate equity and the exercise of corporate control by banks. Bank-oriented systems are thought to display high levels of bank finance, equity holding by banks, long-term relations, close monitoring and active corporate governance by banks.

In practice, the distinction between bank- and market-oriented systems is fragile.[3] While bank lending to corporations has been high in Japan in comparison to the UK and USA, it has not in Germany. Bank holdings of corporate equity are modest in most countries. While banks are thought to have been actively involved in corporate activity and in particular restructurings in Japan, they have not in Germany. In addition, although early studies of Japan pointed to the advantages of close bank–firm relations in Japan, more recent ones have noted their defects in displaying excessive conservatism in corporate lending and inhibiting restructuring.[4]

The influence of financial systems on measures of corporate governance is also unclear. Close relations between financial institutions and companies may have been thought to influence incentives and disciplining of management. Systems with close relations have better information flows and thus a firmer basis on which to reward and discipline management. But they may lack the powerful incentive and disciplining devices of stock markets. In fact, to the extent that there is evidence on this, it does not point to a clear difference in either incentive arrangements or disciplining across financial systems.[5]

3. OWNERSHIP AND CONTROL

The standard bank–market orientation distinction is neither particularly robust nor insightful. In contrast, there are striking differences in the ownership and control of companies that do bear close scrutiny.[6] This is normally discussed in terms of comparisons of concentration of ownership in the UK and USA on the one hand, and Continental Europe and the Far East on the other. For example, in France and Germany, in more than 80 per cent of the largest 170 listed companies, there is a single shareholder owning more than 25 per cent of shares and, in more than 50 per cent of these companies, there is a single majority shareholder. In contrast in the UK, in only 16 per cent of the largest 170 listed companies is there a single shareholder owning more than 25 per cent of shares and in only 6 per cent is there a single majority shareholder. Concentration of ownership is appreciably higher on the Continent of Europe than in the UK. High levels of ownership concentration have also been reported for the Far East and South America and ownership is as dispersed in the USA as in the UK.

Not only does the level of ownership differ appreciably between the UK and USA and most of the rest of the world but so too does the nature of that ownership. In the UK and USA, the shares of listed companies are primarily held by institutions, such as pension funds, life insurance firms and mutual funds, and individual investors. Ownership is dispersed in the sense that no one institution or individual holds a large stake in a single company. This is described as an 'outsider system'.[7]

On the Continent and in the Far East, the large share blocks are primarily held by families (or family holding companies) and other firms. Intercorporate holdings of large blocks of shares are commonplace, frequently in the form of pyramids of shareholdings, cross-shareholdings or complex webs. As noted above, in most countries, bank holdings of shares are modest and holdings by the government vary appreciably across countries. This is described as an 'insider system'.

4. COMPARATIVE INSTITUTIONAL ADVANTAGE

A theoretical literature is emerging suggesting a relation between the institutional structure of countries and the *types* of activities that are undertaken in those countries. There are several strands of theory pointing in this direction. These can be classified under the headings of information, renegotiation and corporate governance. In the information theories (see, for example, Allen, 1993; Allen and Gale, 1999) new technologies, where there are legitimate grounds for diverse expectations, benefit from securities markets. More traditional investments, which are prone to asymmetries of information between borrower and lender, benefit from the economies of monitoring that banks can provide. In the renegotiation theories (see, for example, Dewatripont and Maskin, 1995), fragmented banking systems are associated with short-term investments and concentrated banking systems with long-term investments. Similarly, dispersed ownership systems are associated with high-risk R&D investments and concentrated ownership systems with lower risk, more imitative investments. In the corporate governance theories (see, for example, Burkhart *et al.*, 1997), concentrated ownership is required to provide active governance of firms by investors but might result in excessive interference. Some activities benefit from the active monitoring of management; others are disadvantaged and require dispersed ownership to discourage investor intervention.

All of the above observations and theories therefore suggest a relation between financial systems and the ownership and control of companies and the types of activities that they undertake. As Carlin and Mayer (2001) argue, they suggest that there is an association between the institutional structure of a country and the activities undertaken in that country. They provide a first empirical assessment of this thesis. They examine the relation between growth and investment in 27 industries in 11 OECD countries over the period 1970 to 1995 and the interaction of the institutional structure of the countries and the characteristics of the industries. They find a close relation between growth and investment of different industries in different countries and the interaction of the structure of countries' financial institutions with the dependence of industries on a variety of financial and other inputs. The relation is particularly significant in the case of R&D. Investment in R&D is closely related to the dependence of industries on equity finance and highly skilled labour and is large in countries with good information disclosure, as measured by accounting standards. The relation between R&D and a high level of skills is pronounced, pointing to the significance of human capital in R&D activities.

The case of high-tech and the financing of new economy illustrates how this relation between financial systems, governance arrangements, legal systems and investment and growth might operate. Germany has a large banking system, a two-tier board structure and a civil law code. The USA has a large stock market, a unitary board and a common law system. The rankings of industries by the intensity of patent registrations for Germany (relative to a 12-country average) are almost inversely related to those for the USA. Information technology, semiconductors and biotechnology, for example, are in the top six (of 30) industries by patent registrations for the USA and in the bottom four for Germany. Germany's patent specialization is highest in civil engineering and transport equipment, which are in the bottom three industries for the USA.[8]

The question that this raises is whether the difference in patent activity in the two countries is related to institutional differences between Germany and the USA. Does the concentration of patent activity in 'science based' industries reflect the advantage of, for example, funding these activities through stock markets and does the more production-oriented patenting activity in Germany relate to its highly concentrated ownership and large banking system? The question that this raises is whether there is an association between these differences in technological activity and the structure of countries' financial institutions. A detailed consideration of the way in which high-tech firms are financed and governed provides some evidence on this.

5. THE FINANCING OF HIGH-TECHNOLOGY INDUSTRIES

The Pre-IPO Stage

The development of high-tech firms involves several phases (see Figure 3.1). The first is the seed stage, when a concept has still to be proven and developed. The second is the start-up phase when products are developed and initial marketing takes place. The firm may be a year old or younger at this stage. The third is the early-stage development when the firm is expanding and producing but may well remain unprofitable; it is often less than five years old at this stage. During the fourth stage of expansion it might go public after six months or a year.

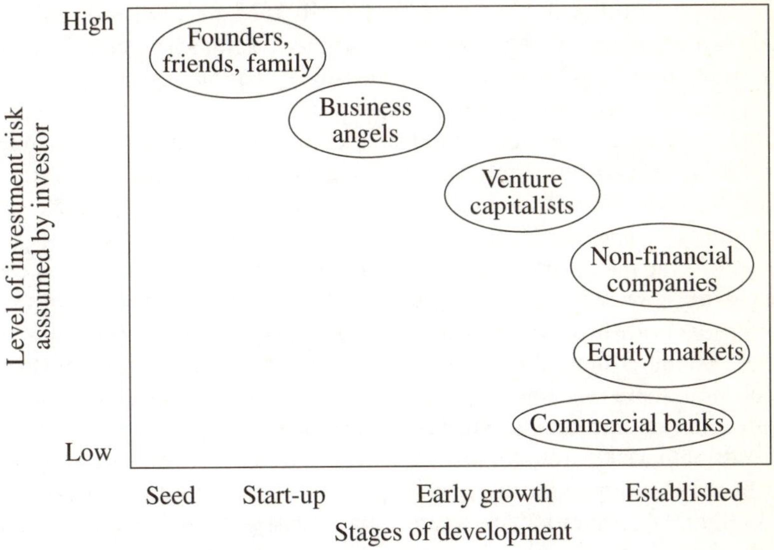

Source: Van Osnabrugge and Robinson (2000).

Figure 3.1 The development and financing of entrepreneurial firms

The initial development almost invariably comes from savings and relatives. Initial external equity financing does not generally come from venture capital firms but from business angels. In the USA, it is estimated that the venture capital industry invested around $5 billion in 1998 in 1000 early-stage firms. In comparison, business angels (wealthy or reasonably wealthy private investors) are estimated to invest $15 billion annually in 60000 early-stage firms. In the UK, it is estimated that about 5 per cent of small firms receive business angel support, as against 1 per cent receiving venture capital finance (quoted in Van Osnabrugge, 1998).

What accounts for the different contribution of business angels and venture capitalists to start-up financing? One of my former doctoral students, Mark van Osnabrugge, undertook a detailed comparison of the way in which venture capitalists and business angels operate. He compared the initial screening, due diligence, investment criteria, contracts, monitoring and exit routes employed by the different types of investor. The results were striking. Venture capitalists are highly rule-based, using careful screening of applicants and due diligence. Business angels place more emphasis on ex post involvement in investments to reduce risks, such as their ability to contribute to the management of the business. Venture capitalists therefore act like institutions following principal–agent relations of limiting risks through monitoring. Indeed, since in the UK they are frequently subsidiaries of institutions, such as pension funds, that is not surprising. Business angels are more actively involved in the subsequent management of activities, exerting more direct control.

From the outset, venture capitalists are focused on exit, business angels much less so. Venture capitalists in general look for rates of return of between 30 and 40 per cent, business angels in the UK between 20 and 30 per cent. Initial public offerings are the preferred route of exit for investors, since they yield the highest return, but they are not the most common. It is estimated that fewer than one in a thousand new ventures have an IPO. However, entrepeneurs are much more optimistic than this record would warrant. One study estimated that 70 per cent of new technology firms believed that a public stock offering was 'highly likely' or 'probable'. Trade sales are the most common exit route of business angels, accounting for over 40 per cent of exits, followed by sales of shares to other shareholders and sales to third parties. IPOs account for just over 10 per cent of business angel exits.

In the USA, around 25 per cent of venture capital funds are invested in early-stage firms. In the UK, start-up and early-stage investments also accounted for around a quarter of venture capital investments in 1984 but this has fallen to a figure of around 4 per cent at present. MBOs (Management Buy-Outs) and MBIs (Management Buy-Ins) have substituted for start-up financing, increasing from 20 per cent to 70 per cent of UK funds' investment.

An important reason for the greater success of US venture capital in funding start-up businesses is the structure of the US industry. Venture capital comprises two parties (see Figure 3.2), the limited partners which are the institutional and individual investors and the general partners which are the venture capital firms investing in individual companies and entrepreneurs. The general partners manage portfolios of companies and are frequently successful entrepreneurs themselves who want to manage larger portfolios of investments. They therefore provide intermediate technical expertise between the investing institutions on the one hand and the entrepreneurs on the other. Venture capital industries in other countries, including the UK, frequently lack the pool of entrepreneurial scientists on which to draw to provide this intermediary function.

Limited Partners

Pension funds

Corporations

Insurance companies ⟶ **General Partners** Venture capital firms ⟶ **Entrepreneurs**

Individuals

Foundations

Foreign investors

Figure 3.2 The structure of the US venture capital industry

The picture that emerges is that the financing of new high-tech firms is highly reliant on own funds, families and friends. Once these are exhausted, external equity initially comes from private investors who are actively involved in the management of the investment. Venture capitalists come in at a later stage, acting more at arm's-length than business angels and seeking higher returns over short periods. A small fraction of the most successful firms are floated on stock markets; most are sold as trade sales and sales to other investors. Much venture capital finance in particular in the UK is not associated with funding new investments but management buy-outs.

To understand high-tech finance, it is therefore important to appreciate it as being intimately connected to the control of firms (Figure 3.3). The transition from personal to business angel to venture capital to stock market finance involves a gradual broadening of the investor base. This moves rapidly from the entrepreneur to single outside investors who are active managers, to financial institutions who use intermediary venture capital firms to screen and manage their investments, to stock markets with largely passive investors.

The financing of Amazon.com illustrates this (see Table 3.1). The firm was initially funded out of Jeff Bezos' own savings and some borrowings. The family then invested a quarter of a million dollars. Two business angels then came in, followed by a larger business angel syndicate. There was a further small family investment, followed by a substantial venture capital injection of $8 million. A year later the firm went public with an IPO of $49 million.

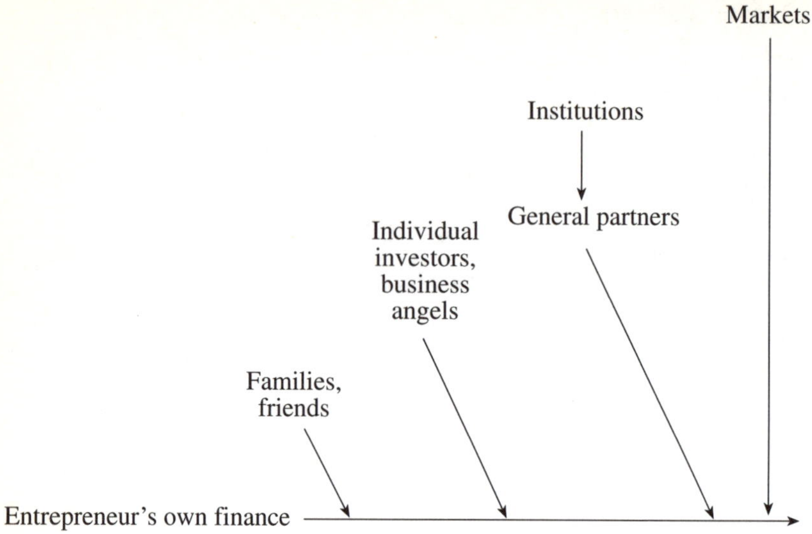

Figure 3.3 Stages of entrepreneurial finance

The Post-IPO Stage

What happens after the IPO? Another former doctoral student of mine, Marc Goergen, has undertaken an interesting comparison of the changing pattern of control of UK and German firms after they have gone public. Goergen notes that historically the average age of a firm coming to the German stock market has been 50 years. In the UK it is around 12 and in the USA around six years. German firms have typically been about twice as large as UK firms on coming to the stock market. At the time of the IPO in general there is either no change in control in Germany, with the original investors retaining control, or control is transferred as a block to a new investor. Even six years after the IPO, families hold majority stakes in nearly 50 per cent of German firms. In the UK, families control a majority of votes in only 11 per cent of firms; most either are taken over or become widely held.

This difference even persists in the Neuer Markt firms. As Vittols (2000) documents, the typical Neuer Markt firm adheres to what is described as the 'Herr im Hause' ('Master of the House') model where the founder/CEO has a controlling stake in the firm and dominates the company board. The innovation strategy is the incremental development of existing products in contrast to that of a venture capital-dominated 'Silicon Valley' firm, which seeks the development of a blockbuster product.

Table 3.1 The financing of Amazon.com (1994–9)

Time line	Price/share ($)	Sources of funds
1994 – July to Nov.	0.001	*Founder:* Jeff Bezos starts Amazon.com with $10000, borrows $44000
1995 – Feb. to July	0.1717	*Family:* Founder's father and mother invest $245500
1995 – Aug. to Dec.	0.1287–0.3333	*Business angels:* 2 angels invest $54408
1995/6 – Dec. to May	0.3333	*Business angels:* 20 angels invest $937000
1996 – May	0.3333	*Family:* Founder's siblings invest $20000
1996 – June	2.3417	*Venture capitalists:* 2 venture capital funds invest $8 million
1997 – May	18	*IPO:* 3 million shares issued, raising $49.1 million
1997/8 – Dec. to May	52.11	*Bond issue:* $326 million bond issue

Source: Smith and Kiholm (2000).

Similarly, in Japan the average age of companies coming to the stock market is significantly greater than in the USA. Sako (2001) reports, for the population of Japanese IPOs in 2000, that the average age of firms coming to the Mothers Market is eight years, 15 years on Nasdaq Japan and 27 years on JASDAQ. The sectoral composition of Japanese IPOs is also quite different from US IPOs. Internet and IT sectors dominate Mothers and Nasdaq Japan, while a large majority of JASDAQ IPOs were in the retail sector.

This further emphasizes the important control differences not only between old and new economy firms but also between different types of new economy firms. There is a much more rapidly changing control structure in new than in old economy firms. Dominant control structures in old economy firms are concentrated and slowly evolving. Dominant control structures of new economy firms shift rapidly between entrepreneurs and different investor groups as the production process and financing needs of firms change.

Examining what happens once firms are established on the stock market further reinforces this observation. Work that I have been doing with Marc Goergen has compared the characteristics of companies listed on the UK stock market with equivalent-sized firms that are privately owned. Consistent with the above observations on the importance of stock markets for high-tech firms, listed firms are concentrated in R&D-intensive sectors of the economy. Listed firms obviously raise much more equity finance but this is not used to fund internal investment. Instead, what clearly distinguishes listed from unlisted firms is the extent to which they engage in acquisitions. Access to stock markets primarily provides firms with the opportunity to expand through acquisition. Stock market listings and dispersed share ownership are important not only in making firms subject to the discipline of the takeover market but in providing them with the opportunity of expanding through acquisitions themselves. Again it is the potential for rapidly evolving patterns of control that mark out the new economy firms.

6. FINANCIAL INSTITUTIONS AND VENTURE CAPITAL FINANCING

We return to the question of the relation of the structure of financial systems to corporate activities in the context of the financing of entrepreneurial firms. To date, very little is known about this. Black and Gilson (1998) have argued that stock markets are a prerequisite to the successful development of a venture capital market and that IPOs provide an important exit route for venture capital funds. But even this assertion is open to question, as a comparison of venture capital in Israel and the UK illustrates. Even though the IPO market has not been active in Israel in recent years, there is widespread investment in firms in their early stages. Instead of using the Tel Aviv market, most high-tech companies go public in the USA on NASDAQ (Blass and Yafeh, 2001). In the UK, where the stock exchange is much larger and more liquid than the Tel Aviv stock exchange, investments in early stages of technological developments are comparatively rare. In addition, according to the venture capital associations, IPOs are nearly as important as an 'exit' mechanism in bank-dominated Germany as they have been in the UK in recent years: 7.5 per cent of all venture capital-backed companies in Germany as against 9 per cent in the UK.[9]

Mayer *et al.* (2001) have undertaken one of the first analyses of the relation between institutional structure and venture capital finance. They examine venture capital industries in four countries: Germany, Japan, Israel and the UK. Their analysis differs from much of the preceding literature in (a) providing an international comparison of countries outside the USA and (b) examining the funds themselves rather than the venture capital firms. The question that the paper poses is, to what extent can differences in venture capital activities (in particular, stages of finance and sector focus) be associated with the venture capital firms sources of finance? Do venture capital firms that are funded through banks invest in firms at different stages of their development from those that are funded by private individuals? Do pension and insurance fund-backed venture capital firms have a different sector focus from corporate-backed funds?

To answer these questions, Mayer *et al.* collect data on venture capital firms and their sources of finance from venture capital associations. The results were striking. Firstly, they report substantial differences across countries in terms of the stage of finance of venture capital firms. In some countries, notably Israel, funds are much more focused on early stage investments than in others, in particular Japan. There is a remarkably close similarity in stage of finance between Germany and the UK, despite the differences in their financial systems noted above.

Secondly, there are significant differences in venture capital firms' sector focus. While biotechnology and life sciences receive a substantial level of attention in all four countries, a much larger fraction of venture capital firms in Israel and Japan invest in information technology, software and electronics than in Germany and the UK, where the manufacturing sector receives more attention.

Thirdly, the paper reports substantial variations in the sources of finance of venture capital firms. Banks are a major source of external finance in all countries, particularly in Germany and Japan.[10] Pension funds are much more significant in the UK than in the other three countries. Corporations are a more important source of finance of venture capital firms in Israel than elsewhere.

Fourthly, the paper reports that there are significant relations between sources of finance of venture capital firms and their investment activities within countries. In particular, banks and pension funds-backed venture capital firms invest in later-stage activities, while venture capital firms relying on private individual investors favour earlier-stage activities. Industry and privately backed funds are focused towards IT, software and electronics and away from manufacturing sectors, while the reverse holds for pension funds.

Fifthly, the paper records significant differences in the relation between financing and investment stage in different countries. While bank-backed venture capital firms in Israel and the UK invest in later-stage activities relative to other sources of finance, bank-backed funds in Germany and Japan are no different from other venture capital funds. Later-stage investing by pension funds is a feature of the UK but not of Israel, the only other country where pension funds are a significant source of venture capital finance.

Institutional differences are therefore associated with significant differences in venture capital activities within countries. But the paper also reports that institutional differences only account for a small proportion of the differences in venture capital activities across countries. This suggests that a majority of international differences are attributable either to demand for funds (that is, supply of entrepreneurs) rather than supply of financial institutions or to the availability of alternative sources of entrepreneurial finance, for example business angels, referred to above. The implication is that, while there may be a matching of institutions with types of entrepreneurial activities within countries, international differences in entrepreneurial activity are primarily driven by other considerations.

7. THE ROLE OF THE STATE

The last few years have seen a rapid rise and collapse in venture capital finance and initial public offerings in many European countries as well as the USA. Bottazzi and Da Rin (2002) report that funds raised in Europe trebled over the two years 1997 to 1999, from €8 billion to €25 billion. Some of the fastest growth has occurred on Continental Europe. Bottazzi and Da Rin (2002) report that, by the end of the 1990s, venture capital investments in Belgium and Sweden represented approximately the same share of GDP as those in the UK, having started from much lower bases at the beginning of the 1990s. Vittols (2000) reports that venture capital in Germany has more than doubled since 1996, from DM6.1 billion to DM13.8 billion in 1999. Early-stage financing has more than doubled as a proportion of total venture capital, from 14 per cent in 1996 to 31 per cent in 1999. Following the establishment of the Neuer Markt in 1997, initial public offerings have increased from 20 in 1996 to 167 in 1999. The recent closure of the Neuer Markt is clearly a significant setback for venture capital financing in Europe.

The growth followed an extended period of abortive attempts to stimulate a venture capital market in Germany. Throughout the 1960s there was a perception that the German economy was suffering from an 'equity gap'. The response of the German government was to encourage the formation of a new institution, the Deutsche Wagnisfinanzierungsgesellschaft (WFG), in 1975, devoted to the financing of young enterprises. In an excellent study, Becker and Hellmann (1999) document the development and performance of the WFG. We summarize below the ownership, investment strategy, performance and subsequent developments of the WFG.

Ownership The WFG was founded by 29 German banks with a government guarantee of 75 per cent of any losses that the banks might incur. The board of the WFG comprised representatives from industry and government, and scientists and consultants.

Investment strategy The WFG initially focused on early-stage investment in particular in manufacturing and information technology. The criteria for selecting investments were the degree of innovation of products and processes, their potential markets, the quality of entrepreneurs and the shortage of alternative sources of finance. The WFG took minority equity stakes, granted the entrepreneur a buy-back right over the equity and took no control rights. It offered entrepreneurs standardized contracts with essentially uniformity of pricing.

Performance The performance of the WFG was disastrous. It made losses in each of its first nine years. Most of the firms it supported recorded net losses at exit and bankruptcies accounted for a high proportion of exits. In contrast to evidence from US venture capital-backed firms, entrepreneurs were ashamed to admit that the WFG had a stake in their firm.

Subsequent developments In 1984, the WFG underwent a fundamental transformation involving a liquidation of the old firm and the creation of a new one. The government exited and the government representatives resigned from the board. The WFG refocused its attention away from start-ups to later-stage financing and shifted its hiring policy away from people with technological experience to those with business experience. In 1988, the new WFG was in turn disbanded and the portfolio was taken over by Deutsche Bank and one small bank.

It is interesting to contrast this case with a similar British organization, the Industrial and Finance Corporation (ICFC).[11]

Ownership ICFC was set up in the UK in 1945 to fill what was known as the 'Macmillan gap' (the failure of the City to supply long-term finance to small and medium-sized firms). It was owned by the UK clearing banks and the Bank of England but created in the face of considerable opposition from the banks, which regarded it as a competitive threat rather than a complementary institution.

Investment strategy ICFC focused on small manufacturing companies in their early development stage. It undertook active screening and monitoring of borrowers. Unusually for British clearing banks, its loan officers had a high degree of technical competence and displayed a high degree of commitment to long-term lending. ICFC took equity stakes but did not have direct representation on the boards of firms.

Performance Following losses in its first three years to 1948, ICFC made substantial profits in every year subsequently. The number of its investments went up by a factor of ten between 1954 and 1984 and investments by ICFC were regarded as signals of quality certification.

Subsequent developments In 1983, ICFC was consolidated in the *3i* group. Its investments become increasingly concentrated on venture capital. Initially it focused on start-ups, early-stage and development capital. By the beginning of the 1990s, it had become by far the largest provider of venture capital in the UK. Increasingly, however, its investments focused on management buy-outs and buy-ins and the repositioning of its activities contributed significantly to the switch in aggregate UK venture capital from early stage in the mid-1980s to MBOs and MBIs by the end of the 1990s.

The markedly different performance of the WFG and ICFC cannot be readily attributed to their origins or ownership structure. Both were formed in response to a financing gap and both had (reluctant) banks as their owners. Neither can the supply of entrepreneurs be cited as a fundamental difference between the UK and Germany. The postwar British culture has been regarded as being particularly anti-commercial. The UK has a creditor-oriented bankruptcy system and a strong bankruptcy stigma. For much of the period there were prohibitively high rates of personal taxation in the UK. However, where the two institutions differed was in the involvement of the banks and the government. In many respects ICFC had the characteristics of a US venture capital partnership: lenders with technical expertise intermediating between investors on one side and companies on the other. The banks were kept at a distance by the presence of the Bank of England and there was no government involvement in ICFC or *3i*. In contrast, the WFG could not disentangle itself from the banks or the government. This suggests that considerable care needs to taken in the design of public sector institutions to promote high-tech finance.

8. POLICY IMPLICATIONS

With the collapse of the Internet bubble and retrenchment of VC firms in the USA and Europe, it might be thought that a discussion of the financing of new technology is largely redundant. But what we are currently experiencing is, of course, precisely what a theory of a relation between institutional structure and corporate activity would predict. Some financial systems are suited to the initial phases of technological innovation that we are currently witnessing and others are suited to the subsequent implementation stages which we are about to observe. Not only are there cross-sectional variations in the relative performance of different systems at a particular point in time but there are also variations in performance of different systems over time.

This is what has been repeatedly observed in the past. The most important periods of stock market expansion have coincided with major technological innovations when returns to investment were exceptionally high. In the UK, these were associated with financing of the canals at the end of the 18th century and investment in railways in the 19th century. But stock markets were less well suited to financing activities that offered more modest returns, most notably investment in manufacturing.

What are the policy implications of this? Becht and Mayer (2000) have recently argued in the context of an analysis of the ownership and control of European corporations that regulation affects the structure of financial and corporate systems. There is evidence that regulatory differences across European countries and between the UK and USA bias institutional arrangements in particular directions.

The regulation of the high-tech sector illustrates this well. While the UK and USA are generally classified under similar common law systems, there are actually pronounced differences between the two countries in their approach to the regulation of non-bank financial institutions, such as pension funds and fund managers. One of the important contributors to the development of venture capital in the USA was the relaxation of the 'prudent man' rule on pension funds at the end of the 1970s. This stimulated a substantial expansion in investment in venture capital activities during the 1980s. US regulation emphasizes the importance of disclosure of information to investors, auditing of the behaviour of institutions and the imposition of penalties, in the event of failure being uncovered.

In the UK, investor protection has relied more heavily on public compensation schemes and the imposition of detailed conduct of business rules. For example, to protect pensioners from the types of losses that were incurred in pension fund scandals during the 1990s, rules were imposed that encouraged pension funds to invest heavily in government securities. These had the effect of discouraging investment in more risky investments such as venture capital funds.

US regulation therefore promotes private contracting, UK regulation relies more heavily on public contracting. Private contracting systems do not require institutions to amass capital before they are allowed to transact. They do not presume that there is a single best way of transacting business and they do not seek to impose common rules of conduct. Instead, they allow institutions and investors to choose how to organize their business and where to invest. If malpractice is uncovered then the intention is that it should be uncovered through auditing and penalized through the courts.

A critical question that this comparison raises is the extent to which reliance should be placed on public versus private contracting to provide protection in non-bank financial institutions. The advantage of private over public contracting is that it does not prejudge what is acceptable. It allows for a greater degree of diversity of institutional form. It permits institutions to adapt more rapidly in the face of changing requirements of both investors and firms. It has therefore made it easier for institutions to respond to the changing financing and control needs of high-technology firms in the USA than in the UK.

On the other hand, it relies on 'caveat emptor' and in general provides investors with less protection than public contracting schemes. It places considerable emphasis on private agents, such as analysts, accountants and auditors, to collect and process information. Recent experience in the USA illustrates very clearly the potential risks and failures associated with this. Furthermore, it relies on the courts to enforce contracts. These are better developed in the USA than elsewhere and it is questionable therefore whether the US model is the appropriate one for other countries

9. CONCLUSIONS

This chapter has argued that there is a close relation between the types of activities undertaken in different countries and their institutional structures. Certain types of institutional arrangement, in particular information disclosure, appear to be related to growth of R&D activities. More generally, there is a relation between the structure of institutions and the types of high-tech activities undertaken. The contrast between German and US patenting and the greater success of the general-limited partnership arrangements in the USA than the captive funds in the UK in funding start-up activities are illustrative of this. So too is the relation between the source of funding of venture capital firms in different countries and the types of activities that they fund.

A distinguishing characteristic of the financing of new economy firms is its evolving pattern of control by different investor groups. Participation in successful firms moves rapidly from own investments, to families, individual investors, small groups of investors and to venture capitalists funded by financial institutions. While stock markets are an important component of the development of the most successful firms, they are not by any means the most common. Where initial public offerings occur, they involve rapid changes in control from original to new investors and dispersed ownership. Stock market finance is important in allowing control of and by high-tech firms to alter.

Regulation is a significant influence on the ability of financial institutions to be able to respond to the changing needs of corporate borrowers. The form in which investor protection is provided affects the degree of risk taking by financial institutions and the types of financing that they offer. This is well illustrated by differences between the public contracting systems of regulating investment management in Europe and private contracting in the USA. Private contracting forms of regulation permit a greater degree of competition and variety of products in financial markets. However, they rely on 'caveat emptor', private firms to undertake monitoring and the courts to enforce contracts. Recent experience suggests the potential hazards associated with this.

NOTES

1. For surveys, see Carlin and Mayer (2000) and Levine (1997).
2. See, for example, Edwards and Fischer (1994).
3. See, for example, Mayer (1988) and Rajan and Zingales (1995).
4. See, for example, Kang and Stulz (2000) and Weinstein and Yafeh (1998).
5. See Kaplan (1994).
6. See Barca and Becht (2001) and La Porta *et al.* (1999).
7. See Franks and Mayer (1995).
8. Patent specialization indices for 30 industries are calculated from patents registered at the European Patent Office. The correlation between the German and US indices is −0.78 (Cusack and Soskice, 2000).
9. EVCA (2000) and BVK (2000).
10. More generally, Sako (2001) notes the relationship between VC firms and old economy firms in Japan, arguing that 'in the future the most successful Japanese incubators will not be affiliated with free-wheeling venture capital funds. Instead, they are likely to be backed by the most forward-looking members of Japan's old mainstream economy: trading companies and banks, manufacturers and consumer companies' (p.12).
11. For an excellent history of the firm and its successor organization, see Coopey and Clarke (1995).

REFERENCES

Allen, F. (1993), 'Stock markets and resource allocation', in C. Mayer and X. Vives (eds), *Capital Markets and Financial Intermediation*, Cambridge: Cambridge University Press.

Allen, F. and D. Gale (1999), 'Diversity of opinion and the financing of new technologies', *Journal of Financial Intermediation*, **8**, 68–89.

Barca, F. and M. Becht (2001), *The Control of Corporate Europe*, Oxford: Oxford University Press.

Becht, M. and C. Mayer (2000), 'The control of corporate Europe', in H. Siebert (ed.), *The World's New Financial Landscape: Challenges for Economic Policies*, New York: Springer.

Becker, R. and T. Hellmann, (1999), 'The attempted genesis of venture capital in Germany: An analysis of the Deutsche Wagnisfinanzierungsgesellschaft', Paper presented at the Corporate Governance Meeting, Venice, 25–26 June.

Black, S. and R. Gilson (1998), 'Venture capital and the structure of capital markets: Banks versus stock markets', *Journal of Financial Economies*, **47**, 243–77.

Blass, A. and Y. Yafeh (2001), 'Vagabond shoes longing to stray: Why foreign firms list in the US', *Journal of Banking and Finance*, **25**, 555–72.

Bottazzi, L. and M. Da Rin (2002), 'Venture capital in Europe and the financing of innovative companies', *Economic Policy*, **17**(1), 229–69.

Burkhart, M., D. Gromb and F. Panunzi (1997), 'Large shareholders, monitoring and the value of the firm', *Quarterly Journal of Economics*, **112**, 693–728.

BVK (2000), *Venture Capital in Europa 1999*, Berlin: German Venture Capital Association.

Carlin, W. and C. Mayer (2000), 'How do financial systems affect economic performance', in X Vives (ed.), *Corporate Governance: Theoretical and Empirical Perspectives*, Cambridge: Cambridge University Press.

Carlin, W. and C. Mayer (2001), 'Finance, investment and growth', mimeo, Saïd Business School, University of Oxford.

Coopey, R. and D. Clarke (1995), *3i: Fifty Years Investing in Industry*, Oxford: Oxford University Press.

Cusack, T. and D. Soskice (2000). 'Patterns of innovation and varieties of capitalism', Berlin: Wissenschaftszentrum.

Dewatripont, M. and E. Maskin (1995), 'Credit efficiency in centralized and decentralized economies', *Review of Economic Studies*, **62**, 541–55.

Edwards, J. and K. Fischer (1994), *Banks, Finance and Investment in Germany*, Cambridge: Cambridge University Press.

EVCA (2000), *Mid-Year Survey, January–June 2000*, European Venture Capital Association (available on the Internet at evca.com).

Franks, J. and C. Mayer (1995), 'Ownership and control', in H. Siebert (ed.), *Trends in Business Organization: Do Participation and Cooperation Increase Competitiveness?*, Tübingen: Mohr Siebeck: reprinted in *Journal of Applied Corporate Finance*, 1997, **9**, 30–45.

Goergen, M. (1998), *Corporate Governance and Financial Performance*, Cheltenham, UK and Lyme, USA: Edward Elgar.

Kang, J-K and R. Stulz (2000), 'Do banking shocks affect firm performance? An analysis of the Japanese experience', *Journal of Business*, **73**, 1–23.

Kaplan, S. (1994), 'Top executive rewards and firm performance: A comparison of Japan and the United States', *Journal of Political Economy*, **102**, 510–46.

La Porta, R., F. Lopez de Silanes and A. Shleifer (1999), 'Corporate ownership around the world', *Journal of Finance*, **54**(2), 471–517.

Levine, R. (1997), 'Financial development and economic growth: Views and agenda', *Journal of Economic Literature*, **35**, 688–726.

Mayer, C. (1988), 'New issues in corporate finance', *European Economic Review*, **32**, 1167–88.

Mayer, C., K. Schoors and Y. Yafeh (2001), 'Sources of funds and investment strategies of venture capital funds: Evidence from Germany, Israel, Japan and the UK', working paper, Oxford Financial Research Centre.

Rajan, R. and L. Zingales (1995), 'What do we know about capital structure? Some evidence from international data', *Journal of Finance*, **50**, 1421–60.

Sako, M. (2001), 'Between Bit Valley and Silicon Valley: Hybrid forms of business governance in the Japanese internet economy', in B. Kogut (ed.), *Global Internet Economy*, Cambridge, MA: MIT Press, pp.291–326.

Smith, R. and J. Kiholm (2000), *Entrepreneurial Finance*, New York: Wiley.

Van Osnabrugge, M. (1998), 'The financing of entrepreneurial firms in the UK: A comparison of business angel and venture capitalist investment procedures', D. Phil, Oxford University.

Van Osnabrugge, M. and R. Robinson (2000), *Angel Investing*, San Francisco: Jossey- Bass.

Vittols, S. (2000), 'Frankfurt's Neuer Market and the IPO explosion: Is Germany on the road to Silicon Valley yet?', mimeo, Wissenschaftszentrum Berlin für Sozialforschung.

Weinstein, D. and Y. Yafeh (1998). 'On the costs of a bank-centred financial system: Evidence from the changing main bank relations in Japan', *Journal of Finance*, **53**, 635–72.

4. The globalization of venture capital: the cases of Taiwan and Japan

Martin Kenney, Kyonghee Han and Shoko Tanaka

INTRODUCTION

At the beginning of the 21st century, the importance of venture capital for the funding of new high-growth potential firms is universally recognized. Many of the defining US firms of the last three decades, including 3Com, Amgen, AMD, Compaq, Cisco, Federal Express, Genentech, Intel, Oracle and Sun Microsystems, were first funded by venture capitalists. Even more than providing funding for firms, venture capital has become a central institution in some of the most dynamic, innovative firm clusters in the world. In the last two decades, venture capital investing has diffused internationally: there are now 35 national venture capital associations. Though the USA continues to be the center of the venture capital industry, there are hot spots of activity in a number of developed and a few less developed nations, nearly all of which are in Asia. This chapter discusses this globalization process and then examines the growth of venture capital in two countries, Japan and Taiwan.

The diversity of nations in terms of their national systems of innovation, levels of entrepreneurship, political economic development, varying labor practices, corporate ownership regulations, educational achievement and business cultures means that each country's venture capital industry has a different evolutionary trajectory. To understand the hybridization of venture capital in different environments, we first construct an ideal type of venture capital drawn from the US experience.

This chapter has seven sections. The first section discusses the history, development and operation of venture capital as an institution and its economic impacts. This is followed by a section that explains what venture capital is and how it operates. The third describes the economic impacts of venture capital. The fourth briefly reviews the globalization of the venture capital industry. The fifth section examines the venture capital industries in Japan and Taiwan. The sixth section summarizes the current situation in the venture capital in light of the severe downturn in global equity markets. The final section summarizes the findings of this chapter.

VENTURE CAPITAL AS AN INSTITUTION[1]

Prior to World War Two, the source of capital for entrepreneurs everywhere was either the government, government-sponsored institutions meant to invest in such ventures, or informal investors (today, termed 'angels') that usually had some prior relationship to the entrepreneur.[2] In general, private banks, quite reasonably, have been unwilling to lend money to newly established firms, because of their high risk and lack of collateral. Venture capital as a formal activity first began on the US East Coast immediately after World War Two. During the next six decades, the practice gradually expanded and became increasingly professionalized, so that today it can be referred to as an industry. By the 1980s, the locus of the venture capital industry had shifted from New York and Boston on the East Coast to Silicon Valley on the West Coast (Florida and Kenney, 1988a, 1988b; Gompers, 1994). Today the ideal–typical venture capital firm is based in Silicon Valley and invests largely in electronics, with lesser sums devoted to the biomedical technologies.[3] Internationally, the largest concentrations of venture capital are in Hong Kong, Israel, London, Taiwan and Tokyo.

The role of the US government in the development of the venture capital industry has been important, but, for the most part, indirect. These indirect impacts include practicing generally sound monetary and fiscal policies, thereby ensuring low inflation with a stable financial environment and currency. Tax policy, though, has been favorable to capital gains, and a number of decreases in capital gains taxes may have had some positive effect on the availability of venture capital (Gompers, 1994). With the exception of a short period in the 1970s, pension funds have been allowed to invest prudent amounts in venture capital funds. The NASDAQ stock market, which has been the exit strategy of choice for venture capitalists, was strictly regulated and characterized by increasing openness (though in light of the recent scandals some of this apparent transparency may have been an illusion). This general macroeconomic environment of transparency and predictability was believed to reduce risks for investors. Put differently, environmental risks stemming from government action and criminal behavior were minimized, though not eliminated.

Another important social policy has been heavy and continuous funding of university research. This produced not only sometimes valuable research results, but also large numbers of graduates with advanced degrees in the sciences and engineering. In the USA, research universities, especially MIT and Stanford, played important roles in the development of venture capital industries in Boston and Silicon Valley, respectively.[4] Biotechnology and the Internet were the direct result of federal funded university research.

The most important direct US government involvement was the passage of the Small Business Investment Act of 1958 authorizing the formation of small business investment corporations (SBICs) for the purpose of investing in small firms of all types. The legislation permitted individuals to form SBICs that were able to borrow money at subsidized rates from the federal government. It also allowed banks through their SBICs to circumvent the Depression Era laws prohibiting commercial banks from owning more than 5 per cent of industrial firms. The final investment format permitted the formation of SBICs that could raise money in the public market.[5]

The SBIC program experienced serious problems from its inception. One problem was that it was very bureaucratic and had constantly changing rules and regulations. Starting in 1965, federal criminal prosecution was necessary to rectify the misappropriation of funds, incompetence and fraud undertaken by some SBICs. By one estimate, 'nine out of ten SBICs had violated agency regulations and dozens of companies had committed criminal acts' (Bean, 2000). Despite the corruption, something valuable also occurred: especially in Silicon Valley, a number of individuals used their SBICs to leverage their personal capital, and some were so successful that they were able to reimburse the program and raise institutional money to become formal venture capitalists. The SBIC program accelerated their capital accumulation and, as important, government regulations made these new venture capitalists professionalize their investment activities.

The most successful case of the export of Silicon Valley-style venture capital practice is Israel, where the government played an important role in encouraging the growth of venture capital (Autler, 2000). The government has a relatively good economic record; there is minimal corruption, massive investment in military electronics research, and an excellent higher educational system. The importance of the relationships between Israelis and Jewish individuals in US high-technology industry for the creation of the Israeli venture capital system should not be underestimated. For example, the well-known US venture capitalist, Fred Adler, began investing in Israeli startups in the early 1970s, and in 1985 was involved in forming the first Israeli venture capital fund (ibid.: 40). Still, the creation of an Israeli venture capital industry would wait until the 1990s, when the government funded an organization, Yozma, to encourage venture capital in Israel. Yozma received $100 million from the Israeli government. It invested $8 million in ten funds that were required to raise another $12 million each from 'a significant foreign partner', presumably an overseas venture capital firm (ibid.: 44). Yozma also retained $20 million to invest itself. These 'sibling' funds were the backbone of a now vibrant community that invested in excess of $3 billion in Israel in 2000, though by the second quarter of 2002 this had fallen to $217 million (PricewaterhouseCoopers, 2002).

In the USA venture capital emerged through an organic trial-and-error process, and the role of the government was limited and contradictory. In Israel, the government played a vital role in a supportive environment in which private sector venture capital had already emerged. The role of government differs. In the USA the most important role of the government was indirect; in Israel, it was largely positive in assisting the growth of venture capital and, in India, the role of the government has had to be proactive in removing barriers (Dossani and Kenney, 2001).

In every nation, the state has played some role in the development of venture capital. Venture capital is a very sensitive institutional form owing to the high-risk nature of its investments, so the state must be careful to ensure its policies do not adversely affect its venture capitalists. Put differently, capricious governmental action injects extra risk into the investment equation. However, in some countries judicious, well-planned government policies to create incentives for private sector involvement have been quite important for establishing an independent self-sustaining venture capital industry.

WHAT IS VENTURE CAPITAL?

Venture capital as a practice is relatively easy to define in the US context, where venture capital and private equity are quite distinct practices, venture capital being the older practice, though one could argue that private equity as a practice is far more like the traditional role of Wall Street financiers, using capital to organize and reorganize firms and industrial sectors. However, for much of the world private equity and venture capital are combined both statistically and in the minds of policy makers. In Europe, a large proportion of what the European Venture Capital Association (EVCA) considers 'venture capital' is, in the USA, considered private equity. So in this section we briefly outline what we take to be venture capital, understanding that this is a relatively narrow definition and that, at the margins, there is ample room for debate.

Venture capital does not come into being and cannot survive in a vacuum, it requires suitable opportunities. The essence of venture capital (and private equity) is the purchase of an equity stake in the venture they are funding. These are not loans though in some cases there are clauses guaranteeing the convertibility of the loans to equity. The success of a venture capitalist is predicated upon realizing the increased value of their equity as the firm grows.[6] Newly established firms have a high mortality rate, so the craft of venture investing is risky. Since these new ventures are very person-intensive and have few fixed assets, so in most failures little can be recovered. Venture capitalists invest in a recently established firm that they believe has the potential to provide a return of ten times or greater in less than five years. They are not interested in funding firms that do not show the potential for a rapid appreciation and they do not evaluate their investments in terms of social goals such as reducing unemployment, increasing R&D or building a community's tax base.

The venture capital process requires that investments be liquidated either through bankruptcy, merger or an initial public stock offering. For this reason, they are only temporary investors and usually are members of the firm's board of directors only until the investment is liquidated.[7] For the venture capitalist, the firm is a product to be sold, not retained. There must be the possibility of moving out from their investments profitably. Nations that erect impediments to any of the exit paths (including bankruptcy) are choosing to handicap the development of venture capital: this is true regardless of the macro-level reasons for the impediment. This does not say such nations will not have entrepreneurship, only that it is unlikely venture capital as an institution will thrive.

In return for investing, the venture capitalists not only receive a significant equity stake in the firm, but they also demand seats on the board of directors from which they intend to monitor the firm. This is important because the venture capitalist intends to provide more than just money, and highlights one of the salient differences between venture capitalists and passive investors: venture capitalists plan to monitor, assist and even intervene actively in their investments. The venture capitalist's objective is to leverage their involvement to increase the investment recipient firm's probability of survival and rapid growth. This involvement extends to the performance of a variety of functions, and can include assistance in recruiting key people, providing strategic advice and introducing the firm to potential customers, strategic partners, later-stage financiers, investment bankers and various other contacts. Experienced venture capitalists, having seen many fledgling firms experience growth pains, are able to provide valuable advice (Florida and Kenney, 1988a, 1988b; Gompers, 1995). It is the venture capitalists' experience, connections and willingness to become involved that differentiate them from other sources of capital.

Thus far no venture capital industry has been able to prosper by liquidating their investments through mergers alone. Jeng and Wells (2000) find that the single best explanation of a vibrant national venture capital industry is the existence of IPO exit strategies. However, this should be qualified by specifying that such exchanges should be liquid and transparent. In other words, stock exchanges that acquire reputations as exits for immature or fraudulent firms will quickly become illiquid as investors refuse to purchase shares in their listings. With such illiquidity come greater opportunities for insider trading and stock manipulation and, barring an effort to reform, an eventual collapse of the exchange. Thus it is not simply the existence of an exchange, but rather a well-disciplined exchange capable of discharging its primary purpose of raising funds for promising companies, that contributes to the growth of venture capital . Finally, it should be understood that the 'exit' of investors and owners is a by-product of the exchange's function of raising capital.

With the exception of Taiwan, the predominant institutional format for venture capital is the venture capital firm operating a series of partnerships called 'funds' investing capital raised from investors consisting of wealthy individuals, pension funds, foundation, endowments and various other institutional sources of funds. The general or managing partners are the professional venture capitalists, while the investors are silent limited partners. The typical fund operates for a set number of years (usually between seven and ten) and then is terminated. Normally, each firm manages more than one fund, simultaneously. The only other persistent source of venture capital has been as subsidiaries of major corporations, that is corporate venturers.

THE ECONOMIC IMPACT OF VENTURE CAPITAL

Measurement of the importance of venture capital in most economies is quite difficult, because in terms of capital investment it is only a minute portion of the total economy. Moreover, the greatest benefits come from Schumpeterian innovations that establish the basis of, not only new firms, but, more important, new industries. Accounting for the economic impact of venture capital is difficult, because it is possible that venture capitalist-backed firms would have come into existence without funding. Entrepreneurs might have funded the firm from other sources or simply boot-strapped the firm by reinvesting retained earnings, though it seems safe to assume that the innovation would have been actualized more slowly.

The anecdotal evidence for the economic importance of venture capital is striking. On the US stock exchanges a number of highly valued firms, including Intel, Cisco and Federal Express, were originally funded by venture capitalists. As of 1999, the US venture capital firm Kleiner, Perkins, Caufield and Byers claimed that the portfolio firms that it had funded since its inception in 1973 had a total market capitalization of $657 billion, revenue of $93 billion, employed 252000 people and had invested in excess of $2 billion (KPCB, 2001). Though extrapolation from KPCB, which is probably the most successful venture capital firm in the world, is risky, it is safe to say that the cumulative impact of the currently over 600 venture capital firms in the USA has been substantial even for an economy as large as the US. In specific regions, especially Silicon Valley and Boston's Route 128, venture capital has been a vital component of what Bahrami and Evans (2000) term 'the entire ecosystem' (see also Lee *et al.*, 2000).

The General Accounting Office (1982: 10) studied the impact of the venture capital industry on the US economy. Extrapolating from 72 publicly listed, venture capital-funded firms in operation in 1979 (there were 1332 venture capital-funded firms in existence at that time), the GAO concluded that employment would increase in 1989 to between 522000 and 2.54 million employees, depending upon the annualized growth assumption. A recent study commissioned by the NVCA (2001) and conducted by the consulting firm WEFA estimated that the firms venture capitalists had invested in were cumulatively responsible for the creation of 4.3 million jobs and $736 billion in annual revenues in 2000.

In the United Kingdom, a survey by the British Venture Capital Association (1999) found that private equity-financed firms grew at an annual compounded rate of 24 per cent, or three times faster than firms in the Financial Times Stock Exchange Index (FTSE) 100 and 70 per cent faster than the FTSE 250. By estimation, they concluded that 2 million Britons or 10 per cent of the current private workforce were employed by venture capital-backed firms. This estimate seems somewhat high, but provides some indication of how important private equity/venture capital has been to the growth of the British economy. In the case of Taiwan, there has been little study of the benefits of the venture capital industry to the entire economy, but many of the most recent Taiwanese computer-related success stories received venture capital funding. The one study attempting to quantify the benefits was done by Wang (1995), who found that the tax deductions encouraged venture capital investments during 1990 to 1992 that were ten or more times greater than the tax dollars expended. In 2000, the Israeli high-technology industry accounted for approximately 25 per cent of the entire GDP, and venture capital investing has been an important support for this high-technology environment.

Another indicator of the significance of venture capital investment is its impact on the innovation process. Kortum and Lerner (2000), using a sample of firms and patent filings, found that venture funding accounted for 8 per cent of US industrial innovations in the decade ended in 1992 and predicted that this could have increased to as much as 14 per cent by 1998. They also found that a dollar of venture capital was 3.1 times as likely to lead to a patent than was a general R&D dollar. Given that venture capitalists, in general, do not invest in process innovations (and these are patented far less frequently), these estimates are likely somewhat high, but they do indicate the importance of venture capital for US innovation. Also this finding might be somewhat of an overestimate, because it is quite likely that some of the inventions venture capitalists are commercializing were actually made in corporate research laboratories. If this is true then the corporate research laboratories would appear to be less efficient than they actually are. However, their result confirmed the importance of venture capital in encouraging R&D investment, and complemented other R&D sources.

There is sufficient evidence to conclude that venture capital has made a significant contribution to the economies of the USA, the UK, Israel and Taiwan. Venture capital backing seems to be an efficient method for commercializing innovations. Though there has been only limited research on the macroeconomic impacts, there is ample evidence that VC has had a significant impact in the USA. It certainly has been the key financier of the US 'New Economy' firms, become a part of the US, Israeli and Taiwanese national system of innovation for commercializing R&D, and become a vital resource in regions such as Silicon Valley and Route 128.

THE GLOBALIZATION OF VENTURE CAPITAL

The genesis of national venture capital industries differs by country. In most of the developed world, the origin was indigenous, both in terms of funding and individual venture capitalists, though the USA was usually the model, whether appropriate or not. The most significant exception is the UK, which had a long merchant banking tradition and formed *3i* immediately after World War Two. In developing countries, international development agencies, especially the International Finance Corporation (IFC), played an important role by encouraging the formation of many of the early venture capital funds, often in cooperation with other donors. For example, a 1986 IFC report identified Malaysia and Korea as candidates for the development of a venture capital industry. The record for the IFC initiatives was mixed, but in cases like Korea and India it did have an important catalytic effect (Dossani and Kenney, 2001). In the early 1990s, the US Agency for International Development also funded a number of venture capital projects, especially in Eastern Europe. One assessment of efforts to encourage venture capital formation in developing nations concluded that in nearly all cases performance was marginal. It found that the return on the IFC venture capital portfolio prior to 1986 was −5 per cent compared to an overall portfolio return of +6 per cent (Fox, 1996). On the other hand, there are cases such as India where there was sufficient success to have justified these investments. Moreover, the Fox study does not recognize other important benefits such as in India where the IFC used venture capital as part of its process in convincing the Indian government to liberalize its financial system (Dossani and Kenney, 2001). Despite these successes, in many other nations such as Brazil, Nigeria, Argentina and Indonesia, bilateral and multilateral efforts to encourage venture investing failed.

The first effort by American firms to export the venture capital model was undertaken by the Rockefeller organization (now known as Venrock), which opened an office in Brussels in 1960, but soon after closed it because of a lack of good investments. Contemporaneously, General Doriot of ARD tried to transfer the venture capital practice overseas and, in 1962, the Canadian Enterprise Development Company (Mather, 2001) and the European Enterprises Development Company were formed (Dominguez, 1974; Hsu, 2002). These two pioneering firms were only marginally successful, and closed in the 1970s (Wilson, 1989). In 1972, ARD also assisted in the creation of an Australian venture capital affiliate (Hsu, 2002).

The first large wave of globalization came in the late 1970s and early 1980s, when a few pioneering US venture capital and private equity firms such as Advent International, Apax, Citicorp Venture Capital and Warburg Pincus established international operations. Europe attracted most of these investments, but owing to the scarcity of start-up investments, very soon the overseas branches emphasized the provision of private equity or expansion investing (Brooke, 2000: 256). Though some of these branches closed during the downturn of the late 1980s, most operations continue and have even expanded.

The movement into Asia was far more fitful and largely came from West Coast firms, though the large East Coast firms did make investments in Asia including Advent International's establishment of Southeast Asia Ventures Inc. in Singapore in 1985. Perhaps the most successful US investor in Asia was Hambrecht and Quist, which began investing in Taiwan in the mid-1980s. However, the boom in western venture capital firms operating in Asia began in the mid-1990s, when the number of firms, the breadth of their Asian operations and their level of activity increased dramatically. The sustainability of these investments in the post-2000 crash period will be sorely tested.

Despite the early efforts of ARD to nurture venture capital in foreign countries, prior to the late 1970s there was little indigenous venture capital overseas. For example, the first French venture capital firm, Sofinnova, established in 1972, began US investing in 1974 through a small joint venture fund established with the assistance of the Boston firm, TA Associates.[8] In 1976, it opened an office in San Francisco to gain access to Silicon Valley deals (Cowley, 2001). By the early 1980s, European banks and financial institutions, attracted by the biotechnology and IT booms, had established venture capital subsidiaries. Because the environment for venture capital investing was far better in the USA, many of these pioneering European subsidiaries established branches in the USA. For example, Dutch bank ING established Atlas Venture, in 1980 and opened its Boston office in 1986 (*Atlas Venture*, 2002). However, in terms of sheer numbers, there was little cross-Atlantic activity.

In Japan, which is discussed in more detail below, there was a similar effort by some of the larger venture capital firms to establish global operations. The most internationalized Japanese venture capital firm is JAFCO (Japan Associated Finance Co., Ltd), which is affiliated to the largest Japanese brokerage firm, Nomura Securities. Established in 1973, JAFCO opened a San Francisco office in 1984, a London office in 1986 and a New York office in 1987. In 1990, it shifted its geographical focus and established an Asian office in Singapore, and has since expanded to eight offices in Asia. The other major Japanese firm with overseas operations is Nippon Investment and Finance Company (NIF). NIF has had a different strategy from JAFCO in that most of its overseas operations are in the form of joint ventures. In 1987, NIF formed a joint venture with the Taiwanese government's development bank. Then, in 1990, it established an office in Singapore, but it was only in 1996 that it opened an office in the USA and started an Israeli joint venture. During the late 1980s, there was concern that Japanese venture capitalists armed with profits from the Bubble would become dominant investors in US high-technology industries. This proved to be unwarranted and now Japanese venture capitalists, both independent and corporate, are minor players internationally.

The most rapidly growing group of venture capital firms investing extranationally are those headquartered in Asia. These transnational venture capitalists see their market as 'Greater China', the area encompassing Singapore, Hong Kong, Taiwan, China and the Chinese in Silicon Valley. (This might be extended to parts of Malaysia and Bangkok, Thailand.) Though there are few statistics to confirm the growth of a pan-Asian Chinese venture capital market, there is anecdotal evidence for its emergence. Curiously, even as the strongest entrepreneurial nation in the region, Taiwan seems to be the least integrated into the pan-Asian venture capital networks.

VENTURE CAPITAL IN TAIWAN AND JAPAN

Taiwan

Taiwan is smaller than Japan, Hong Kong and even Singapore in terms of the available venture capital and yet, in terms of the number of start-ups and the success of its venture capital investors, Taiwan is the most active spot in Asia for venture investing. The reasons for this are multifaceted and relate to a background of entrepreneurship, linkages with the USA, especially Silicon Valley, an early emphasis on electronics as a key industry, a supportive government and a national emphasis on education. These factors created an environment in which industrial growth and venture investing combined into a virtuous circle reinforcing the practices of entrepreneurship and venture investing. The history of the Taiwanese venture capital industry is intertwined with the development of its electronics industry.

The Taiwanese electronics industry

The Taiwanese electronics industry has been built on a combination of steadily increasing manufacturing expertise (this includes product design within well-understood product trajectories) and strong linkages to the US market. In historical terms, it was foreign investors that made it possible for Taiwanese firms to gain access to global markets and absorb new technologies. These factors strategically positioned local entrepreneurs within international markets.

Taiwan developed a tool kit of manufacturing expertise that was a direct complement to the tendency in Silicon Valley to outsource various production activities. Moreover, the target industries were experiencing extremely rapid growth. This set of capabilities, when combined with an entrepreneurial mentality, meant that Taiwan was able to advance far beyond its initial strategy of leveraging low-cost labor. These conditions were the ones into which the government launched its efforts to create and support a venture capital community.

Not surprisingly, with such an industrial backdrop, Taiwanese venture capital industry is the most Silicon Valley-like in Asia. In terms Florida and Kenney (1988a) used, it would be considered a 'technology-oriented' complex. This can be seen through the aggregate investment statistics. The concentration of investment in electronics and information technology is striking when compared to, for example, Japan (see Table 4.1). This pattern was noticed even at the birth of the Taiwanese venture capital industry as the *Venture Capital Journal* (1985: 3) wrote: 'investments will be made in products that have been tested in the marketplace'. The earliest venture investments were in firms proposing to undercut Japanese firms supplying either US firms on an original equipment manufacturer (OEM) basis or the US market directly on the basis of price. Presciently, the *Venture Capital Journal* concluded that venture investing would be concentrated in the computer industry, an observation that remains true today. Taiwanese venture capital firms had two investment strategies, one for their Taiwanese investments and one for US investments. In Taiwan, investments were concentrated in firms able to use their skills at decreasing production costs to transform a high-cost, cutting-edge product into a commodity. In the USA, the investments were usually in firms that had a Chinese founder or co-founder, and that might be able to utilize Taiwanese production capabilities in some way.

Table 4.1 Investment fields of Taiwanese and Japanese venture capitalists, 2000 (per cent)

Industry	Taiwan	Japan
Agriculture/Fisheries	—	0.5
Computer-related	20.2	8.7
Conglomerates	2.1	0.8
Construction	0.6	2.0
Consumer products/services	3.2	9.4
Electronics	17.5	9.8
Ecology	1.0	0.2
Financial services	3.0	27.8
Information technology	17.6	7.1
Infrastructure	2.0	1.7
Leisure/Entertainment	2.2	0.7
Manufacturing – heavy	7.6	10.9
Media	2.3	0.6
Medical/Biotechnology	5.5	2.0
Mining and metals	—	0.7
Retail/Wholesale	1.2	0.8
Services, non-financial	1.8	4.3
Telecommunications	9.5	9.9
Textiles and clothing	0.5	0.6
Transportation/Distribution	1.0	0.8
Travel/Hospitality	0.9	0.3
Utilities	0.3	0.4

Source: Calculated by the author from *Asian Venture Capital Journal* (2002).

Technical and industrial expertise is important; however, as important are the sociocultural institutions that encourage entrepreneurial activity. Taiwan has a long history of extended families investing in entrepreneurial ventures that have some sort of family linkage (Hamilton and Biggart, 1988), though the venture capital business was not a direct descendant of these types of linkages. If family-funded entrepreneurship was an enabling condition, curiously it also created obstacles. For example, many foreign venture capitalists have criticized Taiwanese entrepreneurs for their reluctance to part with equity and to subject themselves to the discipline required by professional investors (Hsu, 1999). Venture capitalists could invest, but they were often limited to small amounts of equity, thus limiting their role in the firm and minimizing their potential gain. Thus the enabling conditions were not sufficient to give rise to a viable venture capital business *sui generis*. There were environmental conditions from which a fledgling venture capital industry could draw, but they did not guarantee that venture capital would emerge spontaneously.

The inception of the industry can be traced to a study trip to the USA and Japan by Li-Teh Hsu, then Taiwanese Finance Minister and the prior Finance Minister K.T. Li. After this trip, which included visits to Japan's Tsukuba, Silicon Valley and Boston's Route 128, they decided to create incentives for the establishment of venture capital in Taiwan (Shih, 1996: 282).[9] Significantly, K.T. Li was quoted as saying one key piece of advice was given to him by Frederick Terman the Stanford University Provost and one of the individuals most responsible for the creation of Silicon Valley: 'lure your talented expatriate engineers home, just as Silicon Valley once lured engineers back from the East Coast' (Knight Ridder News Service, 1999).

In 1983, legislation was passed giving attractive tax incentives to individuals willing to invest in professional venture capital firms. The most important feature was an up to 20 per cent tax deduction for Taiwanese individuals provided they maintained their venture capital investment for at least two years. To qualify, the investment had to be made by a venture capital fund approved by the Ministry of Finance and in a Taiwanese firm or a foreign firm that would transfer technology to Taiwan. Notice investing abroad was acceptable as long as a benefit to Taiwan could be shown. This was very important because it allowed Taiwanese venture capitalists to forge strong ties with Silicon Valley. Each investment in a firm would be examined by an auditor who decided what proportion met the government criteria of 'high technology' and then a percentage rebate of up to 20 per cent was approved. An investor receiving the full deduction only risked $0.80 for every $1.00 of investment. Whereas the initial law stated that only individuals could receive the rebate, the statute was revised in 1991 to allow corporate investors the same 20 per cent tax deduction. This revision dramatically increased the amount of venture capital. The ultimate indicator of the program's success came in 1999, when the government declared that the venture capital industry was mature and discontinued the 20 per cent tax deduction program.

The tax deduction was by far the largest benefit, but there were others. For example, 80 per cent of the investment income was exempt in the current fiscal year, providing a grace period of one year. Also those who chose to reinvest the earnings garnered from a venture capital investment were allowed to deduct the venture capital income from their tax return in that year (Asian Technology Information Program, 1998; Republic of China, 1996: 9–10). This encouraged the investors to allow the various venture capital firms to reinvest their earnings. The Taiwanese government also undertook other measures to ensure the growth of venture capital. One measure was a willingness to invest government funds in venture capital firms provided that they were matched by those from the private investors. In return for these incentives, there were restrictions as to which industries were eligible for investment, and the government excluded investments in publicly traded securities, real estate and retail operations. Thus there was a quid pro quo: in return for the various benefits the government could restrict and channel the activities of its venture capitalists.

Among the earliest venture capital firms to join the scheme were two USA-based operations, one of which was the basis for the creation of the pan-Asian firm H&Q AP. However, for these early venture capitalists, despite the attractive benefits, there were few attractive opportunities and returns were marginal until 1990 (Schive, 1999: 102; Asian Technology Information Program, 1998). And yet, despite the weak returns, the number of venture capitalists increased. However, after 1990, the average returns were positive (Schive, 1999). During the earlier period, it was successful investments in Silicon Valley and the potent subsidy program that guaranteed the survival of the venture capitalists. The government subsidy programs appear to have been very successful because, according to the calculations of Wang (1995: 89), the multiplier effects of the government's use of tax deductions to encourage venture capital were 'ten-fold or above' in the years 1990 to 1992.

The environment improved from 1994 onwards: the Taiwanese stock market rose and the profitability of the venture capitalists increased. The reasons for the increased profitability are threefold. First, Taiwan experienced the 'New Economy' boom and then a speculative fever fueled the rise of technology stocks as global IT spending soared. Also competition in the computer business, especially personal computers, became ever more intense, prompting US firms to outsource their production to Taiwan. Second, the 1991 rule change permitting operating companies to receive the tax deduction for investing in venture capital firms encouraged an increasing flow of capital. Third, Taiwanese technological capabilities improved and the linkages with US firms strengthened.

The globalization of the Taiwanese venture capital industry

The Taiwanese venture capital community was born globalized. Every important Taiwanese venture capital fund has an office in the San Francisco Bay Area, and invests between 20 and 30 per cent of its total funds outside Taiwan (*Asian Venture Capital Journal*, 2001). According to Taiwanese venture capitalists that I interviewed in 2001, this was almost exclusively directed to Silicon Valley. An indicator of the destination for these investments is the geographic preferences listed by the firms in the AVCJ. The USA is overwhelmingly the preferred location for overseas investing as nearly 40 per cent of the Taiwanese venture capital funds mentioned a willingness to invest in the USA (see Table 4.2). Since 2000, Taiwanese venture capitalists have begun investing more actively in the Shanghai area, in parallel with the establishment of branch manufacturing operations in this region.

*Table 4.2 Investment preferences of Taiwanese venture capitalists**

Country (n = 7)	Number stating preference (n = 71)**
Taiwan	30
No preference***	41
United States	29
China	9
Singapore	6
Malaysia	4
Thailand	4
Hong Kong	4

Notes:
* This only includes firms headquartered in Taiwan.
** More than one nation was possible.
*** We assume that those registering no preference are probably limiting their preferences to Taiwan. In fact, those stating no preference were the smaller funds.

Source: Calculated by the author from *Asian Venture Capital Journal* (2001).

The other aspect of globalization is USA-based firms investing in Taiwan. From the aggregate statistics, this appears to be limited. However, two important venture capital firms in Taiwan are H&Q Asia Pacific and Walden International Investment Group. These two firms were early investors in Taiwan and they continue to be significant. More recent US entrants, such as WI Harper and Crimson Ventures, also have offices in Taiwan and are significant investors. For example, Asian Americans or Asian immigrants largely staff these USA-headquartered firms and most of the investors are wealthy Asian individuals or firms (Hellman, 1998). More recently, large US financial institutions have also begun to invest in Asia-centered funds. Finally, the largest Japanese venture capital firms including JAFCO, NIF, JAIC and a few others have offices in Taiwan.

In 2002, the Taiwanese venture capital industry faced a variety of challenges, some of which are related to the internal Taiwanese situation and some of which are imposed by macroeconomic developments. How the government and the venture capital firms respond will have significant impacts upon the health of the venture capital industry and their continuing ability to support Taiwanese start-ups. The most significant immediate challenge is the elimination of the 20 per cent tax deduction. This removed a significant investment incentive, though in 2000 it did not appear to decrease fund raising dramatically. However, after the weak performance of the Taiwanese stock market in 2000, in 2001 very few funds could be raised (Yang, 2001). If the stock market continues in such a depressed condition, the IPO window will remain closed and it will be difficult for venture capitalists to raise new funds, especially as the risk reduction generated by the tax deduction will be eliminated. For the many small venture capital firms, this could cause closure and/or distress mergers. The other challenge is whether and when to invest in China, which is rapidly attracting manufacturing and even R&D from Taiwan.

Today Taiwan has the most dynamic technology-driven venture capital industry in Asia. Outside of the USA, the only consistently more successful venture capital industry is Israel. From this perspective, there is no doubt that the decision in the 1980s to subsidize the creation of a venture capital industry was a major policy triumph. Experienced professionals with good track records have created a vibrant industry. However, the next few years will be a stern test of the industry's ability to survive. Without the deduction, the decisions by investors will be driven solely by returns. The government has stated that the venture capital industry is mature, and yet it has not deregulated the industry to allow it to operate as though it is mature. The Taiwanese government will have to make some difficult decisions regarding deregulation of this 'mature' industry.

Japan

Since the collapse of the 'bubble economy' in 1989, the Japanese economy has stagnated. Once seemingly invincible financial institutions are currently burdened with non-performing loans. Despite these difficulties, in terms of technological R&D capacity, especially in electronics, Japan continues to lead Asia and is second only to the USA. Japanese universities, though on average not as strong as elite US universities in terms of research, clearly have pockets of world-class engineering and scientific research excellence and the graduates are quite capable. In the private sector, the long-term employment system creates significant obstacles to labor mobility especially among the most highly trained and talented. Thus it seems fair to say that Japan is the Asian nation with the greatest human resources capable of being the raw material for an entrepreneurial economy, but just as clearly the entire socioeconomic system is not organized to encourage high-technology entrepreneurship.

Japan has a long history of entrepreneurship and a significant small business sector that can be traced back into the Tokugawa Shogunate, when the regional samurai rulers encouraged enterprises in their fiefdoms in an effort to capture income. Immediately after World War Two, there was a phase of intense entrepreneurship and many new firms such as Sony, Honda and Alps were formed. Local, prefectural and national government agencies also had various loan programs for small and medium-sized enterprises (SMEs). Japanese policy makers have actively encouraged and protected SMEs (for a description, see Nishiguchi, 1994). However, until the 1990s, this interest, for the most part, did not extend to support for start-up firms intent upon entering new markets.

The history of venture capital in Japan[10]

The first effort to develop venture capital came in 1963, when the Japanese government authorized the use of public funds from the national and prefectural governments and the private sector to establish three small and medium business investment and consultation companies in Tokyo, Nagoya and Osaka. This program differed from the US SBIC program in that in Japan only three firms were formed, whereas in the USA, by 1963, nearly 500 SBICs had been established. The majority of equity in these three firms was held by local governments, city and regional banks, insurance firms, stock exchanges, private corporations, and chambers of commerce. Up to March 1996, these three Japanese SBICs had cumulatively invested 69.2 billion yen (at an average conversion rate of 150 yen to the dollar, this is in excess of $400 million) in 2500 firms, of which 78 undertook public stock offerings. Though they played an important role in supporting existing SMEs by providing stable, long-term capital, they invested in very few start-ups (Niimi and Okina, 1995).

In the early 1970s, Japan experienced its first venture capital boom. In 1972, the first private venture capital corporation, the Kyoto Enterprise Development company (KED) was established with investments by 43 prominent local firms, including Kyoto Stock Exchange, Bank of Kyoto, Tateishi Electric and Industrial Bank of Japan. The motive force behind KED was the Kyoto Association of Corporate Executives aiming to promote knowledge-intensive industries as a regional development strategy. The model for KED was the first US venture capital firm, American Research and Development. However, KED was unsuccessful and was liquidated only four years later (Ono, 1995). Also, in 1972, the Nippon

Enterprise Development (NED) was formed by a group of 39 firms, and included both financial institutions and venture businesses (NED was liquidated in 1999). In 1973, Japan Godo Finance, which was the precursor to the present JAFCO, was established by Nomura Securities and 15 other shareholders. In total, between 1972 and 1974, eight private venture capital firms were formed by major banks, such as Sumitomo, Mitsubishi, Daiichi Kangyo, and security firms, such as Yamaichi and Nikko. In other words, the major Japanese financial institutions formed venture capital subsidiaries. Most of the personnel were seconded from the investors for a few years and then returned to their original organizations, thereby limiting the creation of seasoned venture capital professionals. Curiously, when one compares the commitment by investors to these venture capital operations to the commitment of US institutional investors, what is striking is that the Japanese are remarkable patient, whereas US corporate venturers usually retreat at the first sign of difficulty. However, Americans who joined the venture capital subsidiaries of US firms often resigned to establish or join an independent venture capital firm, while Japanese professionals who served in the venture capital subsidiary nearly always returned to the parent firm. In this way, the American individuals did exhibit a long-term commitment to venture capital, even though the institutions did not. The ultimate result was that the USA accumulated a corps of trained venture capitalists, while in Japan this corps formed far more slowly.

The 1973 oil crisis recession ended the first expansion phase. The number of investments declined and the industry stagnated. However, of the eight firms formed, six survive until this day. So a venture capital industry was created, but it did not have sufficient success to become an important part of the Japanese political economy. Thus the firms remained small divisions in large Japanese financial institutions.

The US hot issues market in the early and mid-1980s, once again, attracted Japanese attention and stimulated a second venture capital boom. The major players in this boom were the security firms and regional banks. Their goal was to use venture investing to create relationships with the SMEs. Their venture capital affiliates provided funding and underwriting to the SMEs with the goal of gaining access to these firms to provide other services. From 1982 to 1984, 37 new venture firms were formed. As in the case of the earlier venture capital firms formed by the largest financial institutions nearly a decade earlier, the purpose of these subsidiaries was to establish relationships with rapidly growing regional SMEs for the purpose of providing other services. Often the regional bank operations were established in cooperation with the major venture capitalists and security firms. As in the case of the first generation of venture capital firms, nearly all of the investments were loans. The Japanese venture capitalists were not seeking capital gains, they had an ulterior motive: they wanted to develop long-term banking relationships with the firms they funded. For this reason due diligence was not so rigorous, because they lent to established firms, not start-ups.

In the second boom the government recognized the need for vehicles for venture capitalists to move out of their investments and organized an over-the-counter stock market. Also, in 1982, JAFCO introduced the first limited partnership investment fund (Hamada 1999: 38–41). However, the second venture capital boom declined as a result of the recession caused by the rise of the yen in 1986 and 1987 after the Plaza Accord. Once again, investment activity declined substantially as few new firms were formed or funded and willingness to lend money to new firms also declined. Then, in 1989, Japan entered into the recession that continues today.

In 1994, and roughly paralleling the growth of the Internet and the upswing in the Silicon Valley economy, Japanese interest in the role of venture capital in facilitating new business formation and the support of start-ups was renewed. This time, however, the boom occurred in an environment in which Japanese industrial and government leaders were far more concerned about the continuing stagnation of the economy. To facilitate new business creation and start-ups in knowledge-intensive and high-technology industries, the Japanese government implemented a variety of new measures. For example, the 1995 Revision of the Law on Temporary Measures to Facilitate Specific New Businesses and the enactment of the Small and Medium Size Enterprise Creation Law in 1995 made SMEs eligible to receive financial as well as informational support. These new laws also encouraged the formation of more venture capital firms and subsidiaries. For example, regional banks and corporations established venture capital affiliates, and some independent firms were formed. For example, in 1996, Nippon Venture Capital was established, with a capitalization of 10 million yen through investments from 41 companies, including Nippon Life Insurance and Ushio Electric. Moreover, corporate venture capital operations such as those of Softbank and Hikari Tsushin began investing.

Operational characteristics

Though independent partnerships are used in Japan, the majority of the venture capital organizations are corporate subsidiaries. Many of the partnerships are also, in fact, operated as corporate subsidiaries. One reason for the paucity of independent partnerships is the relative lack of available institutional funds. The largest source of funds in the USA, pension funds, is forbidden from investing in the risky area of venture capital. Additionally, until recently, Japanese investors were subject to unlimited liability, making investment risky. According to a 1997 survey, 63 per cent of the total venture capital investments were made by the venture capital subsidiaries, while the remaining 37 per cent came from partnerships. The use of the partnership mechanism increased, especially after the passage of the 1988 Investment Operations Responsibility Association Law that limited the investor liability. This has been especially noticeable recently, as a survey by the Venture Enterprise Center found that the number of venture capital partnership funds increased from 174 in 1999 to 238 in 2000 and they now account for 47 per cent of the total investment (METI, 2001: 8).

The source of funds for Japanese venture capitalists has largely been financial institutions and domestic operating companies, though JAFCO operates a number of relatively small funds with investments from Japanese institutions. With pension funds forbidden from investing in venture capital, and a general lack of university endowments and large tax-exempt foundations, institutional investors simply were not a significant funding source. Among venture capitalists, the affiliates of security firms obtain funds from a variety of sources, and invest through partnerships. The bank and insurance-affiliated venture capitalists get investment capital from their parent firms in the form of debt, rather than through an equity investment.

One of the defining characteristic of the Japanese venture capital industry is that loans are its preferred form of capital disbursement. The reasons for the use of loans by Japanese venture capitalists can be traced to characteristics of both venture capitalists and the start-up firms. Since Japanese venture capitalists receive their capital through loans, they are required to pay interest. The difficulty is that equity investment assumes that for some period there will be no return, thus forcing the venture capitalist to repay its loan out of its initial capital – a difficult requirement. Quite naturally, this encouraged venture capitalists to provide loans, but, of course, this limited their upside potential so that they needed to find low risk opportunities. One solution was to structure the loan with convertibility or a large equity kicker, but the entrepreneur often balks at paying interest and losing equity. This situation is even less tractable because of the long time it takes for Japanese firms to reach an IPO. The average age of firms at the time of their IPO is more than ten years. This means that Japanese venture capitalists must support a firm longer and cannot liquidate their investments as quickly as their US counterparts.

Japanese venture capitalists traditionally invest in the later stages of a firm's growth. According to a 1995 survey by the newspaper *Nihon Keizai Shimbun*, the percentage of companies receiving investment that were 20 years old or older was 48.7 per cent in 1994 and 35 per cent in 1995 (Ono, 1997: Ch.7). This may be changing as earlier-stage investments increased dramatically after 1995. For example, a METI survey conducted in 2000 found that the share of investment in portfolio companies less than five years old was 62 per cent in FY 1999, an increase of 50 per cent on 1998.

There are difficulties with Japanese entrepreneurs also. As in many other nations, Japanese entrepreneurs often have the goal of creating a 'family' firm, so they are reluctant to cede large equity interest to other investors. With this goal in mind, they often prefer loans and bonds. The situation for venture capitalists is complicated further by the Anti-Monopoly Law prohibiting any single investor (including venture capitalists) from owning more than 49 per cent of the total equity and when the shareholding is greater than 25 per cent, the shareholder is not allowed to control the board of directors. For this reason, Japanese venture capitalists usually acquire no more than 20 per cent of their portfolio companies' equity, resulting in relatively small investments. For example, in 2000, the disbursements per portfolio company for initial and follow-up rounds were $500 000 for corporate venture capitalists and approximately $700 000 for independent venture capital firms (METI, 2001: 9). This means that the typical investment is small and the normal monitoring that is so much a part of the value-added of a venture capital investment is economically infeasible owing to the large number of small investments.

The Japanese venture capital investments are largely domestic (70 to 80 per cent of all new investments). Another 20 to 30 per cent was committed overseas, with Asian firms receiving between 10 and 15 per cent of the total, while another 5 to 10 per cent was invested in North America.

This lack of an economic incentive, combined with the lack of experience on the part of Japanese venture capitalists, means that they are relatively uninvolved with their portfolio firms. This is especially important when it comes to assisting their portfolio firms. The lack of in-house skills and experience means that they can perform only cursory monitoring of their firms and are incapable of providing advice and assistance based on experience. The result is that Japanese venture capitalists are largely passive. In this sense, their relationship with their portfolio firms resembles that of a banker, thus explaining the emphasis on loans.

The Internet bubble and beyond

At the time, the Internet boom appeared to be a defining moment for the Japanese venture capital industry as it led to a rapid growth in entrepreneurship. It energized young entrepreneurs and venture capitalists and contributed to what has been called the 'Bit Valley' phenomenon.[11] For example, in 2000 there were approximately 1300 new Internet-related companies located in 23 wards of Tokyo, about 40 per cent of which were in the Shibuya and Minato Wards.[12] In fact, one-fourth of those in Tokyo (371 in 2000) are clustered around Bit Valley (Aoyama, Harajuku, Shibuya and Ebisu). These Internet-related firms are generally small firms and relatively new: 39 per cent employed 30 employees or less and 49 per cent were founded after 1994. Many of entrepreneurs who started the Internet business gained their education and/or work experience in the USA and have established links to American Internet companies in Silicon Valley and the Silicon Alley (Arai, 2000). Though no hard figures are available, by 2002 Bit Valley had experienced a very painful shake-out paralleling the one experienced by Silicon Alley in New York.

In response to the excellent market for IPOs, venture capitalists began making equity investments and it appeared for a moment that an equity-based start-up culture was emerging. Leading this change was Masayoshi Son's Softbank, which had been enormously successful, investing in US Internet start-ups such as Yahoo!, Geocities and E*Trade in the 1995–7 period. It rapidly globalized its investment activities but, most important for Japan, Softbank began a massive venture investing program in Japan. Softbank soon became one of Japan's foremost venture capital firms, and by 2001 in terms of accumulated total investment ranked second only after JAFCO. Softbank was emulated by others, one of the most important of which was Hikari Tsushin, a corporate venture capitalist that is today close to bankruptcy. The boom also encouraged the existing venture capitalists that had been loan-oriented to shift to equity. In addition, a number of Internet firm incubators, such as Neoteny and Netyear Knowledge Capital, were formed on the premise that they would invest in the seed and start-up stage of Internet firms. These organizations differed from the traditional venture capitalists, because they were independent and dedicated to early-stage investment. The Internet boom did prompt a dramatic shift in the perspective and outlook for venture capital in Japan.

The collapse of the Internet bubble in the USA triggered a global collapse, which Japanese venture capitalists have not been able to avoid. Softbank, for example, announced a 890 billion yen loss ($740 million at 120 yen = $1) for FY 2002, and is liquidating many of its holdings (Softbank, 2002). Hikari Tsushin, owing to bad investments and management, hovers close to bankruptcy. In FY 2002, JAFCO's earnings dropped by 60 per cent (JAFCO, 2002). Though the most recent statistics for 2000 do not indicate a drop in commitments, there can be little doubt that they have plummeted. At this point the long-term future of an equity-based venture capital industry in Japan is uncertain. In August 2002, NASDAQ Japan announced that it was closing down altogether.

The role of government and regulations

The core of Japanese regulations for venture capital and SMEs is the Japan Small and Medium Enterprise Agency (SMEA), an affiliated institution of the Ministry of Economy, Trade and Industry (METI, formerly the

Ministry of Trade and Industry), and MITI (METI), which has continuously developed policies to promote SMEs in general. However, they have not been specifically concerned with start-ups and venture capital until recently. Only in the 1990s, with the low rate of new business formations, did the SMEA and METI begin to place more emphasis on new firm formation and venture capital as a source of innovation and employment.

Since the mid-1990s a series of legislations have been enacted to support new firms funded by venture capital. For example, in 1994 the Fair Trade Commission amended its regulations to permit venture capitalists to serve on their portfolio firms' board of directors. Also the 1989 Law on Temporary Measures to Facilitate Specific New Business was revised in 1995 to extend financial and informational support as well as loan guarantees to firms qualifying as 'venture' firms, that is, those firms producing a new product or service or using a new technology to enhance their existing products or services. Also revised was the Commercial Code allowing firms to issue stock options, something that had for all intents and purposes been prohibited. This allowed venture business firms to begin using stock options as an incentive for employees and board members. With the new regulations, venture business firms that qualified for specific METI programs could have option pools of up to 30 per cent of their outstanding shares, whereas other firms were not allowed to issue options for more than 10 per cent of the issued shares.

Another major change in 1998 was the enactment of the Limited Partnership Act for Venture Capital Investment. Prior to the passing of this law, all the investors in the partnership funds had to assume unlimited joint liability. With the new law, the regulations governing investment in partnerships were the same as those in the USA. The liability of investors was limited to their original investment as long as it met METI's official criteria. Also, in 1997, in a measure aimed at stimulating angel investment, the Japanese government introduced a regulatory change, the so-called 'Angel Tax' allowing investors to deduct their capital losses from capital gains on other investments.

During the last decade, the Japanese government has developed policies aimed at supporting new start-ups and removed many of the legal and regulatory obstacles to the practice of venture capital. In regard to venture capital itself, it has not created significant programs to provide incentives to increase the amount of venture capital available, probably, in large measure, because Japan has a surfeit of venture capital. One unusual feature of the Japanese venture capital scene is that Japan is the only significant advanced developed country that did not have a national venture capital association until 2002, remarkably late for a country like Japan, where policy is often driven by industry associations.

The globalization of venture capital in Japan

For foreign venture capitalists, entry into Japan has proved difficult and often not so profitable. The usual strategy has been to form a joint venture with a Japanese financial firm. Moreover, as in Europe, venture capital and private equity are not strictly delineated, and most foreign investors are said to be more interested in private equity firm restructurings than in start-ups (AVCJ, 2002). However, in early 2000, a METI-VEC survey showed the dramatic increase as foreign investors contributed 26.3 per cent (177.9 billion yen) of the total investment in new venture capital funds established between July 1999 and June 2000 (METI, 2001: 27). The US venture capitalists included both corporate venture capitalists, such as Intel, and other institutional investors, such as Goldman Sachs and GE Capital. For example, J.H. Whitney & Co. committed approximately $200 million; Apax/Patricof raised nearly $180 million in cooperation with the Japanese firm, Globis, and Schroeder Ventures secured approximately $150 million (Netry.com, 2000). GE Capital Corp. launched an approximately $180 million fund in conjunction with Daiwa Securities. A $270 million fund was established by the Goldman Sachs Group and Kyocera Corp for the express purpose of investing in high-tech firms (Spindle, 2000). In 2002, the attitude of foreign firms to venture capital investment in Japan became far less optimistic.

As mentioned earlier, Japanese venture capital firms began their international activities quite early. There were two prime target areas. The first target was the USA, simply because it had the most opportunities and the best deals. In the mid-1980s, there were fears that the Japanese would 'buy' Silicon Valley, but this fear proved unfounded as many of their investments failed dramatically. Somewhat later in the 1980s, Japanese venture capitalists also began operations in Asia. Here there were two important types of investments: first, they could provide loans to smaller Japanese firms establishing operations in Asia; second, they could provide funds to Asian firms that had contracts to supply Japanese manufacturing operations in the region.

The most globalized Japanese venture capital firm is Softbank, whose initial success came from the successes mentioned earlier. Building upon these investments, Softbank swiftly expanded to become a global investor. In the heady days of the 1990s, it created subsidiaries in the USA, UK, Continental Europe, Latin America, China and Korea. The remit of these subsidiaries was to make venture capital investments in joint ventures and local Internet start-ups. In 2000, Softbank and the IFC launched the Softbank Emerging Markets Fund. This and other Softbank–IFC joint ventures, worth approximately $520 million, were meant to combine Softbank's Internet and new firm formation expertise with the IFC's experience to fund developing country firms, encourage entrepreneurship and build up the developing nations' Internet infrastructure. The result of this initiative is not known; however it provides insight into how expansive the Softbank vision was. The leading Japanese venture capital firms have been quite active globally, but, as Table 4.3 indicates, most firms are focused on Japan.

Summary

Despite the size of the industry in terms of capital, it is not a significant aspect of the economy or the national system of innovation. In large measure, this is due to the existing bank-oriented financial structure and a thicket of government regulations that have only recently been changing. This has made it difficult to develop a dynamic venture capital industry along the lines of the ones in the USA, Taiwan or Israel. The bank-oriented system has lacked institutions, such as a developed stock market for equity in new firms and an incentive structure aligned with the needs of entrepreneurs. Given the relative underdevelopment of the independent venture capitalists and the previous lack of stock markets or acquisitions as an exit possibility, it is not surprising that traditional conservatism of the corporate venture capitalists dominated the Japanese venture capital scene (Saijo, 2000: 26–9).

The industrial structure of the Japanese venture capital industry still poses problems for the development of a vibrant industry. This is illustrated in the way the venture capital subsidiaries of financial institutions that dominate the system reflect the interests of their parent organization rather than those of the independent venture capitalists. The Japanese scene continues to have many 'venture capitalists' that do not actually have any experience as venture capitalists. Moreover, most Japanese venture capitalists do not have their personal interests linked to the success of venture capital as an institution.

In the last five years, there has been a significant change in governmental policies and regulations. Moreover, there is increased public awareness of new firm formation and the need for a viable venture capital industry. The effectiveness of public policies and resources encouraging more entrepreneurship remains uncertain.

The arrival of the Internet economy and opening of new stock markets, especially MOTHERS and NASDAQ Japan (though in August 2000 NASDAQ Japan announced it was closing), increased the opportunities for entrepreneurs to launch new businesses and raise funds in the early stage of businesses. The increased exit opportunities allowed venture capitalists to be more active in taking risks in promising venture businesses in their early stages. Further, the influence of Softbank's Masayoshi Son in raising

*Table 4.3 Investment preferences of Japanese venture capital firms**

Country (n = number of firms)	Number stating preference (n = 147)**
Japan	110
No preference***	37
United States	25
Singapore	14
Taiwan	10
Hong Kong	8
Thailand	8
Malaysia	7
Asia	5
Philippines	5
Europe	4
Indonesia	4
India	3
China	2
Australia	2
Other (mentioned by one)	6

Notes:
* This only includes firms headquartered in Taiwan.
** More than one nation was possible.
*** We assume that those registering no preference are probably limiting their preferences to Taiwan. In fact, those stating no preference were the smaller funds.

Source: Calculated by the author from *Asian Venture Capital Journal* (2001).

awareness among government officials, industrialists and, most important, entrepreneurs should not be underestimated. This effect may survive beyond the current depressed market conditions.

This climate shifted drastically in 2001 and the situation for both the new firms, independent venture capitalists and operations such as Softbank became decidedly negative. Management buy-outs and other such private equity-based strategies will be more important than pure venture capital investing (*Nihon Keizai Shimbun*, 2001), though in 2002 the market became so difficult that even this strategy may no longer be viable. Despite the promising developments at the end of the 1990s, the current difficult market, the general shortage of experienced venture capitalists, the low levels of labor mobility, and general risk-averse tendencies for the entire society, the development of a Silicon Valley-like venture capital industry is unlikely in the next five years.

THE CURRENT SITUATION

The globalization of the venture capital industry has occurred in two ways: First, in at least 35 nations we were able identify there is now an independent national venture capital association that invariably contains national venture capitalists. In some other nations, such as Sri Lanka, there is not yet a venture capital association, though there are a few venture capitalists. During the 1990s, fueled by the overheated stock markets, the desire to have a venture capital industry became a global fad with each nation and often subnational entity attempting to establish a venture capital industry. The hype and apparent solidity of this boom induced many nations to establish new stock markets with loose listing requirements that attracted dubious and even fraudulent firms. The unfortunate reality is that this was a mania and, as Kindleberger (1978) so effectively described, the consequences of this mania began unfolding in late March 2000 as the market for these firms crumbled.[13] As a consequence, the IPO window closed and by the end of 2000 large corporations also suffering from the downturn were no longer willing to acquire start-ups at inflated prices or, in fact, at any price.

For venture capitalists, 2000 was difficult and 2001 would be even more severe. In 2002, many of the largest and most successful US venture capital funds decided to release their limited partners from some agreed-upon capital calls. In 2002, more money was returned to investors than was raised as the US industry decreased its size, better to reflect the new reality. Smaller and newer venture capitalists were forced to leave the business entirely. Corporate venture capitalists that had been so active in the USA and globally retreated. The cyclical nature of the venture capital industry reasserted itself. In the global context, this raises the question of whether venture capitalists in the smaller markets will survive. In Brazil, for example, the venture capital market has collapsed almost entirely, though there still is a private equity market. This will probably continue until at least 2004.

GENERAL OBSERVATIONS AND CONCLUSIONS

This study indicates that the successful transplantation of venture capital as an institution is predicated upon a receptive environment. Most important, of course, are the business opportunities that permit large capital gains. One of the greatest difficulties in examining Japanese venture capital is to decide whether Japan really has a venture capital industry. By our strict initial definition, it would be possible to argue that most of what is considered venture capital in Japan is not venture capital at all, and oddly enough, since it was loan-oriented it was not private equity either.[14] In the case of Japan, the environment was not conducive and opportunities for investments in firms capable of large capital gains were largely unavailable except during the Internet bubble. In contrast, the Taiwanese environment was more conducive, thus providing opportunities for venture capitalists.

Governments clearly have a role to play in the creation of a venture capital industry; however, as we saw in Japan, even government support is not sufficient to establish an industry if the other features of the environment are negative. In Taiwan, a strong venture capital industry grew with government suppor in a local environment that was conducive to entrepreneurship. Now the Taiwanese government has discontinued its most powerful tool for promoting the industry, so the next few years will be a severe test. This chapter also suggests that, if there are an insufficient number of entrepreneurs or a lack of attractive business sectors, government incentives will only temporarily boost entrepreneurship. So, in the nations that have been most successful in creating venture capital industries, there was government assistance, and this was probably a necessary element for starting venture capital growth, but clearly not sufficient.

Government regulations can also retard the development of venture capital. For example, in the case of Japan, regulations prevented venture capitalists from undertaking their monitoring and control functions. This created moral hazards and prevented the evolution of a healthy relationship between the entrepreneur and the venture capitalist.

We also found that banks find it difficult to perform the venture capital function. While it is not necessary to forbid bank investment in venture capital, they probably should not receive incentives. The USA has always had loose bankruptcy laws, and it might be beneficial for nations with stricter rules to revise them to be less punitive and to work at removing the stigma of failure, thereby lowering the social barrier to entrepreneurship. In some nations, cultural and social changes may need to occur. One possible strategy for this would be an effort to shift the society's estimation of entrepreneurship.

Today there are globalized venture capital firms, such as 3i, Apax, Atlas Ventures and Vertex Ventures (from Singapore), but, for all intents and purposes, the markets they operate in are national. The most important exception to this is Europe where the European Venture Capital Association represents its members in Brussels. This is creating a pan-Europe market, encouraging venture capitalists to syndicate deals outside their home nations, most of which are too small to have a sufficient number of deals to support a venture capital firm. There are also a few firms, such as Walden, WI Harper and H&Q Asia Pacific that operate across Asia and the US West Coast. A number of these firms consider this as part of a 'Greater China' strategy. However, in Asia the national venture capital industries remain separated by languages, customs and laws. The final form of globalization is the bilateral relationships that have evolved between Silicon Valley and Israel, Taiwan and India.

There can be little doubt that in the last decade venture capital has diffused to many more nations. Despite this diffusion, in many of these nations venture capital has not yet made a significant impact. This could be due to its relatively recent establishment or because the new environment inhibits growth. It is also possible that these national venture capital industries will hybridize to better operate in their host nations. This might explain why, in the more bank-centric Continental Europe, venture capital industries are largely private equity-oriented and do less start-up and early-stage investing. In Israel, a nation with high education levels and a relatively entrepreneurial population, but an almost non-existent internal market, a strong venture capital industry focusing upon early-stage investing has emerged. The Taiwanese venture capital industry specializes in electronics manufacturing and chip design, fields where it has national advantages. The Indian venture capital industry focuses upon services such as software and business process outsourcing. The environment within which they operate and are embedded, shapes the character of national industries.

The US venture capital model has been a remarkable success over the last 50 years and contributed significantly to the nation's development. However, it may not be model for all nations. And yet, the Taiwanese and Israeli experience indicate venture capital can provide important assistance to the growth of high-technology firms and industries. Moreover, it is also important to note that venture capital and high-technology entrepreneurship are not the only ways to develop, as Japan and many other developed nations have shown.

ACKNOWLEDGMENTS

Martin Kenney bears responsibility for all errors and opinions. He would like to thank Yili Liu and Tze-chien Kao for their assistance in understanding venture capital in Taiwan.

NOTES

1. This section is adapted from Kenney and von Burg (1999) and Dossani and Kenney (2001).
2. On angels, see Robinson and van Osnabrugge (2000).
3. There are, of course, important venture capital firms headquartered in other regions, and there is a diversity of venture capital specialists. For example, there are funds that spe-

cialize in retail ventures. Some of the largest venture capital funds such as Oak Investment Partners and New Enterprise Associates have partners devoted to retail ventures, though their main focus is IT.

4. Kuemmerle (2001) argues that the private universities were the reason for the success of the US venture capital industry. This underestimates the role of UC Berkeley in Silicon Valley and UCSF in the Bay Area biotechnology industry. This certainly is not a general rule globally as, for example, publicly funded institutions in Israel and Taiwan were critical.
5. Publicly held venture capital funds have a remarkable record of failure. Nearly every other publicly owned venture capital fund has failed. The only important exception is the very first publicly held venture capital fund, American Research and Development, which provided investors with a reasonable but not outstanding return. This is not only true for the USA after repeated waves of start-ups, but is also true overseas. For example, nearly all of the 'incubators' offered on the UK AIM market have now failed.
6. Unfortunately, the value can also grow owing to the belief by others that the investment has become more valuable, even though there may not have been a true growth in the firm's performance. Thus hype about an investment area such as biotechnology, the Internet or nanotechnology can assist the venture capitalist in securing a large capital gain. This is, of course, a region where venture capital can approach fraud, though the investment bank's desire to protect their reputation is supposed to prevent the offering of truly fraudulent firms to the public.
7. This is not always true. Arthur Rock, the lead venture capitalist in funding Intel, remained on the Intel Board of Directors for two decades. Donald Valentine, the lead venture capitalist in funding Cisco, continues on the board fully a decade after it went public.
8. Advent International spun out of TA Associates.
9. Stan Shih, the founder and chairman of Acer, accompanied them on this trip (Shih, 1996: 282).
10. See, also, Kuemmerle (2001) for a discussion of the Japanese venture capital industry.
11. Bit Valley is taken from Shibuya (which literally translated means 'bitter valley'), where many Internet-related firms were established.
12. This description of Bit Valley draws upon Yukawa (2000).
13. Kenney (2003) describes the roots of the US Internet bubble. My argument is that, despite the billions of dollars that US investors squandered on the Internet, in the long run the benefit to the US economy will be that its firms such as eBay, Amazon, Google and Yahoo! will ultimately be globally dominant thanks to the first-mover advantages. In the long-term this should be an important benefit to the US economy, though the short-term pain for US investors, especially those that purchased Internet firms at initial public offerings, is enormous. Notice that the wisest venture capitalists had cashed in many of their investments and thus experienced fewer losses.
14. Oddly enough, in Europe private equity is considered part of the venture capital industry, while in Japan venture capital investing is in the form of loans.

REFERENCES

Arai, Hisashi (2000), *Bitto Barei no Kodo*, Tokyo: Nikkei BP Shuppan.

Asian Technology Information Program (1998), *Venture Capital in Taiwan*, 2 February.

Asian Venture Capital Journal (AVCJ) (2001), 'The 2001 Guide to Venture Capital in Asia', Hong Kong: AVCJ.

Asian Venture Capital Journal (AVCJ) (2002), 'The 2002 Guide to Venture Capital in Asia', Hong Kong: AVCJ.

Atlas Venture (2002), website accessed on 8 September.
Autler, Gerald (2000), 'Global Networks in High Technology: The Silicon Valley-Israel Connection', Master's thesis, Department of City and Regional Planning, University of California, Berkeley.
Bahrami, Homa and Stuart Evans (2000), 'Flexible Re-Cycling and High-Technology Entrepreneurship', in M. Kenney (ed.), *Understanding Silicon Valley: The Anatomy of an Innovative Region*, Stanford: Stanford University Press, pp.165–89.
Bean, Alden (2000), 'Invisible People: Race, Gender, and the SBA in the 1950s', unpublished manuscript.
British Venture Capital Association (1999), 'The Economic Impact of Venture Capital in the UK', London: BVCA.
Brooke, Peter (2000), 'Peter Brooke – Advent International', in U. Gupta (ed.), *Done Deals*, Cambridge, MA: Harvard Business Press, pp.245–58.
Chung, Yulanda (2000), 'World's Worst Bourse?', *AsiaWeek* 11 August.
Cowley, Louise (2001), 'Founding Fathers', *European Venture Capital Journal* (1 May) (http://www.ventureeconomics.com/evcj/protected/ctryreps/ZZZJOXHLBZC.html).
Dominguez, John R. (1974), *Venture Capital*, Lexington, MA: Lexington Books.
Dossani, Rafiq and Martin Kenney (2001), 'Creating an Environment: Developing Venture Capital in India', unpublished manuscript, 27 February.
Florida, Richard and Martin Kenney (1988a), 'Venture Capital-Financed Innovation and Technological Change in the US', *Research Policy*, **17**(3), 119–37.
Florida, Richard and Martin Kenney (1988b), 'Venture Capital, High Technology and Regional Development', *Regional Studies*, **22**(1), 33–48.
Fox, James W. (1996), 'The Venture Capital Mirage', US AID Program and Operations Assessment Report 17.
General Accounting Office (1982), 'Government–Industry Cooperation Can Enhance the Venture Capital Process', 12 August, GAO/AFMD-82-35.
Gompers, Paul (1994), 'The Rise and Fall of Venture Capital', *Business and Economic History*, **23**(2), 1–26.
Gompers, Paul (1995), 'Optimal Investment, Monitoring, and the Staging of Venture Capital', *Journal of Finance* **50**, 1461–89.
Hamada, Yasuyuki (1999), *Nihon no Bencha Kyapitaru*, Tokyo: Nihon Keizai Shimbun.
Hamilton, Gary and Nicole Woolsey Biggart (1988), 'Market, Culture and Authority: A Comparative Analysis of Management and Organization in the Far East', *American Journal of Sociology Supplement*, **94**, S52–S94.
Hellman, Thomas (1998), 'WI Harper International: Bridge between Silicon Valley and Asia', Stanford University, Graduate School of Business, SM-39, 13 February.
Hsu, David (2002), 'The Evolution of Organizational Practices at an Early Venture Capital Firm: A Study of American Research and Development, 1946–1973', unpublished manuscript, August.
Hsu, Ta-Lin (1999), *Asian Venture Capital Journal*, December, 26.
JAFCO (2002), www.jafco.com (site accessed 6 September, 2002).
Jeng, Leslie A. and Philippe C. Wells (2000), 'The Determinants of Venture Capital Funding: Evidence Across Countries', *Journal of Corporate Finance*, **6**, 241–89.
Kenney, M. (2003), 'The Growth and Development of the Internet in the United States', in B. Kogut (ed.), *The Global Internet Economy*, Boston: MIT Press, pp.69–108.

Kenney, Martin and Urs von Burg (2000), 'Institutions and Economies: Creating Silicon Valley', in M. Kenney (ed.), *Understanding Silicon Valley: Anatomy of an Entrepreneurial Region*, Stanford: Stanford University Press, pp.219–40.

Kindleberger, Charles Poor (1978), *Manias, panics, and crashes: a history of financial crises*, New York: Basic Books.

Knight Ridder News Service (1999), 'Taiwan's Silicon Valley Is Booming', 21 September, http://www.mercurycenter.com/asia/center/hsinch091999.htm.

Kortum, S. and J. Lerner (2000), 'Assessing the Contribution of Venture Capital to Innovation', *RAND Journal of Economics,* **31**(4), 674–92.

KPCB (Kleiner, Perkins, Caufield & Byers) (2001), www.kpcb.com.

Kuemmerle, Walter (2001), 'Comparing Catalysts of Change: Evolution and Institutional Differences in the Venture Capital Industry in the US, Japan and Germany', unpublished paper, Harvard Business School, 12 June.

Lee, Chong-Moon, William Miller, Marguerite Gong Hancock and Henry Rowen (2000), 'The Silicon Valley Habitat', in C-M. Lee *et al.* (eds), *The Silicon Valley Edge*, Stanford: Stanford University Press, pp.1–15.

Mather, Derek (2001) 'Personal Interview by Martin Kenney', 6 April.

Ministry of Economy, Trade and Industry (METI) (2001), *Annual Survey of Japanese Venture Capital Investments*, prepared by Venture Enterprise Center (VEC), Tokyo: VEC.

National Venture Capital Association (NVCA) (2001), 'New Study Documents 4.3 Million Jobs and $736 Billion in Annual Revenues Created by Venture Capital Investments', 2 May.

Netry.com (2000), 'Venture Capital Directory', http://www.netry.com.

Nihon Keizai Shimbun (2001), www.nikkei.co.jp/news/tento, 5 June.

Niimi, Kazumasa and Yuri Okina (1995), 'Bencha Bijinesu no seicho o Habamumono ha Nanika', *Japan Research Review*, May, http://www.jri.co.jp/jrr/1995/199505/.

Nishiguchi, Toshihiro (1994), *Strategic Industrial Sourcing: The Japanese Advantage*, New York: Oxford University Press.

Ono, Masato (1995), 'Venture Capital in Japan: Current Overview', November, http://www.asahi-net.or.jp/~sh3m-on/venture capitalommune/javc/jvcs.htm.

Ono, Masato (1997), *Bencha Kigyo to Toshi no Jissaichishiki* (Knowledge of Venture Business and Investment), Tokyo: Toyokeizai Shimpo-sha.

PricewaterhouseCoopers (2002), 'The Kesselman and Kesselman Pricewaterhouse-Coopers Money Tree Survey for the Second Quarter of 2002'.

Republic of China, Ministry of Finance (1996), *The Venture Capital Industry in the Republic of China*, May.

Robinson, Robert J. and Mark van Osnabrugge (2000), *Angel Investing: Matching Startup Funds with Startup Companies*, San Francisco: Jossey-Bass.

Saijo, Nobuhiro (2000), 'Wagakuni Shokenshijo-kan no Genryu', *Shoken Keizai Kenkyu*, **24**, 19–33.

Schive, Chi (1999), 'How Did Small and Medium Enterprises in Taiwan Survive the Crisis?', 1999 Industry Economics Conference Proceedings, Monash University, 12–13 July, 91–110.

Shih, Stan (1996), *Me-too Is Not My Style*, Taipei: Acer Foundation.

Softbank (2002), www.softbank.co.jp/en/newsrelese/2002release/e020426_2.htm (6 September).

Spindle, Bill (2000), 'Japan Becomes Mecca for Venture Capitalists', *Wall Street Journal*, 22 May, http://www.startup.wsj.com.

Venture Capital Journal (1985), 'Venture Capital Debuts in Taiwan', *Venture Capital Journal*, January, 2–4.

Wang, Lee-rong (1995), 'Taiwan's Venture Capital: Policies and Impacts', *Journal of Industry Studies*, **2**(1), 83–94.

Wilson, John (1989), *The New Venturers,* Reading, MA: Addison-Wesley Publishing Company.

Yang, Theresa (Deputy Secretary General, Taiwan Venture Capital Association) (2001), 'Personal Interview by Martin Kenney', 30 April).

Yukawa, Kou (2000), 'Tokyo ni okeru net-kigyo no shuseki, Nihon-ban sirikon arei no hatten ni mukete', Kenkyu Report no.88, Fujitsu Research Institute.

5. Targeting venture capital: lessons from Israel's Yozma program

Gil Avnimelech and Morris Teubal

INTRODUCTION

This chapter deals with policies promoting venture capital, particularly, although not exclusively, targeted incentives programs aimed at VC emergence, such as Israel's Yozma program (successfully implemented during 1993–7). Its focus is policy issues suggested by the Israeli experience. In Israel, the VC industry did not arise in a vacuum; rather it evolved from a prior setting of high-tech and R&D/innovation capabilities. In contexts of this kind Israel's experience (or parts of it) may be relevant. This experience also suggests that a VC-directed incentives program should not represent the central thrust of government policy aimed at creating a completely new high-tech sector (Gelvan and Teubal, 1997). The chapter builds upon prior earlier work on Israel's VC and high-tech industries. It also links with a long tradition of research on VC covering both 'positive' and 'normative' aspects.

Background Research on Venture Capital

The literature of the 1980s, such as Florida and Kenney (1988a, 1988b) focuses on the roles that VC played in the innovation process of the USA. VCs are active investors and are integrally involved in the creation of start-up companies; they are involved in four overlapping networks of innovation: financial institutions, local and global technology markets, professional business service markets and professional labor markets.

During the 1990s we observe at least two strands of literature. The first analyzes how the operation of VC, its mechanisms and organizational capabilities, helped overcome the 'lemon' problem (Akerlof, 1970; Myers and Majluf, 1984) and other information-related problems (Stiglitz and Weiss, 1981) associated with the financing of high-tech start-up companies (for summaries, see Gompers and Lerner, 1999, 2001). A second strand analyzes the impact of VC on their portfolio companies' success. For example, Florida and Smith (1994) found the VC-backed start-up companies are more global than non-VC-backed firms while several studies show that the presence of a VC in an issuing firm serves to lower total cost of issuing: it reduces IPO underpricing and also underwriters' cost. Other studies show that VC serves to lower bank interest rates on loans while also enabling younger firms to go public (Megginson and Weiss, 1991; Barry, 1990; Gompers and Lerner, 1999). It was also shown that VC-backed IPOs have better post-IPO performances in terms of both stock price and growth rates (Megginson and Weiss, 1991) compared to non-VC-backed start-ups. Finally, Megginson and Weiss (1991) compare the post-investment evolution of VC-backed firms with non-VC-backed firms and find that the VC-backed firms have higher growth in terms of total assets and revenues, and invest a larger fraction of total expenses in R&D. Research on the Israeli VC industry showed similar results: VC-backed firms have superior performance compared with non-VC-backed firms, including higher exit rate, younger age at IPO, higher IPO valuation and higher growth in sales (Ber, 2002; Lukomet, 2001; Avnimelech, 2002). More recently Kortum and Lerner (2000) found that VC in the USA spurs technological innovation both among the firms receiving the financing and within the entire sector. According to their paper, on average each dollar invested by venture capital contributes to the rate of patents three to four times more than corporate R&D. Moreover, from the late 1970s to the mid-1990s, VC represented only 3 per cent of corporate R&D, but was responsible for 10 per cent–12 per cent of privately funded innovation.

VC policy literature

All in all, the above literature is testimony that significant progress has been made in understanding the operation and impact of VC. This contrasts with research on VC *policy*, which has been much less extensive and much less focused and successful in generating new knowledge. It is our belief that part of the problem resides in the policies implemented themselves – their simplistic underlying assumptions and their widespread failure – both of which explain why no satisfactory conceptual framework has yet been developed. It also explains why we believe that Israel's successful Yozma program could trigger the creation of such a policy framework.

Public policy aimed at stimulating venture capital was significant in the early-to-mid-1980s, when most of the European countries implemented VC-directed policies. Most focused on the VC supply side: how to increase the pool of VC capital, through reduction of capital gain taxes, tax benefits, preferred loans and government guarantees (OECD, 1997). Very little attention was given to measures that stimulated the demand for VC, such as the establishment of new start-up companies, both in terms of quantity and of quality. Moreover, with few exceptions, no serious attention was paid to measures to attract professional high-quality VC managers and firms into the VC industry. One exception is Poterba (1989) who examines whether and how capital gain taxation influenced the growth of the VC industry. He examines both the supply side (pool of capital to VC funds) and the demand side (the motivation of individuals to become entrepreneurs and to join start-up firms). However, even in his work, no attention is given to the creation of a pool of knowledgeable, professional VC managers despite the strong supply inelasticities in their generation (Gompers and Lerner, 1999, ch. 1).

By the late 1980s and early 1990s the scant success of such programs led to a disappointment with VC policies in general. Again a precursor was Florida who argued that government programs aimed at developing national VC industries failed (Florida *et al.*, 1990; Florida and Smith, 1994). This failure was related to the following facts: VC investments flow mainly to established high-tech centers regardless of the geographical location of the VC industry, a fact which means that it has a weak impact in regions without established high-tech clusters.[1] It follows that, for VC-directed public policy to be successful, the VC industry support must be part of a broader and more comprehensive set of policies which supports the whole high-tech cluster (R&D, innovation, start-ups and exit). These conclusions have been confirmed by other research claiming that government policies are not effective in the creation of a successful VC industry. Presumably these views caused most researchers in the 1990s to ignore policy issues in the field of the VC industry.

The interest in public policy for VC rose again in the aftermath of the enormous success of the VC industry in a number of countries during the 1990s and a few successful government policies supporting and triggering VC industries (these included the Yozma program in Israel). Black and Gilson (1998) emphasize the interaction between the strength of the local IPO market and the development of the VC industry. Jeng and Wells (2000) show that 'the initial public offering market does not seem to influence commitments to early stage funds as much as later stage ones'. In general this strand of research suggests that the strength of the local IPO market is mainly related to late stage VC investment while the supply of high-quality start-up firms is more significant in early stage VC investments. Lerner (1999) and Gans and Stern (2000) examined the success of the SBIR program in the USA (a significant program supporting start-up companies in the USA). Other research tries to explain why government VC programs succeeded in some countries and failed in others.

Despite the above and despite the fact that some recent policy-related research in the field takes into consideration both VC demand and VC supply, most of the existing research still ignores issues of organization: how to attract professionals to the VC industry and how to stimulate cumulative learning and VC capabilities generation. The orientation up to now has focused on incentives (particularly supply incentives) rather than how to develop a high-quality, professional, VC industry.

Recent Research on Israel's Venture Capital Industry

In recent work, Avnimelech and Teubal (2002b) analyzed the emergence and development of a venture capital industry in Israel and its role in the recent successful growth of Israel's high-tech cluster. Taking an evolutionary and systemic perspective, we traced the co-evolutionary and dynamic process involving the business sector, technology policies, venture capitalists, individuals and start-up companies, and foreign linkages. VC emergence is part and parcel of the reconfiguration (Teubal and Andersen, 2000) of a pre-existing electronics industry, one involving large numbers of start-ups and new and powerful links with global capital markets. The main conclusions and policy lessons of the paper are that specific technology policies aimed at the venture capital sector can be effective in causing VC emergence only to the extent that (a) favorable background conditions exist or are created;[2] (b) a pre-emergence period existed with a significant amount of informal VC and start-up related activities; and (c) the design and timing of such policies was such that they led to the early and rapid build-up of reputation and capabilities.[3] The Israeli experience will form the basis of a conceptual framework for the evaluation of VC policies (Avnimelech and Teubal, 2000d).

In Israel, *background conditions* included a pre-existing high-tech industry which developed considerable innovation capabilities during a 10–20 year period[4] – the result of a coherent and important horizontal program supporting company R&D;[5] significant restructuring of the pre-existing military dominated electronics industry during the second half of the 1980s; domestic stabilization policies and capital market liberalization; and globalization of technology capital markets (NASDAQ). Also business links with US industry and capital markets (spurred by the BIRD program which promoted joint R&D between Israeli and USA companies[6]) were being established. Moreover, during pre-emergence (1989–92) a considerable number of business experiments took place, both with respect to the structuring of a new type of start-up oriented both to product and to capital markets (with some success stories) and also in relation to VC-related activities. There was also important policy experimentation and learning for example from the launch of the relatively unsuccessful Inbal program in 1992 and from the technological incubators program. This, together with the rate of start-up creation at the time and large-scale failure in implementing R&D results, suggested the existence of a 'systemic' failure in Israel's business sector. Policy makers eventually agreed on the means to overcome this deficiency-creation of a domestic VC industry with a dominant limited partnership (LP) form of organization.

The pre-emergence conditions specified above enabled an appropriate design of a targeted VC policy program (Yozma) which stimulated VC entry of professional VC companies and 'collective' learning. These and other factors spurred a self-reinforcing, cumulative process of VC emergence and development. VC–start-up coevolution, which parallels supply–demand and user–producer links and interactions in young markets, represented one important axis and distinctive characteristic of the cumulative process that took place in Israel. Other distinctive aspects are the large numbers of IPOs in global markets, the large scope of merger and acquisition (M&A) activity which took place) and the favorable world product and capital market conditions (at least until 2000).[7]

We conclude that the emergence of Israel's venture capital industry could be visualized as a path-dependent process involving a broad set of economic, societal and even geopolitical factors (some endogenous and some exogenous) spanning two to three decades.

Venture Capital and High Tech Development During the 1990s

Increasingly during the 1990s, the evolution of venture capital has been linked to the evolution of high-technology, although there are significant differences among countries in this respect.

Broadly speaking, the Israeli high-tech experience of the 1990s is seemingly quite similar to that of Silicon Valley (both during emergence and during the 'reconfiguration' of the 1980s growth in the 1990s (see Saxenian, 1998, ch.5). The main difference concerns policy: despite numerous US government programs supporting small companies and also R&D (see Lerner, 1999) and despite the role of the SBIC program there was no background 'backbone' program which parallels the role of Israel's horizontal R&D grants scheme.[8] Moreover, despite the importance of the US's SBIC program for the subsequent emergence of venture capital, there was no specific policy aimed at creating a proper venture capital industry (despite some general policies in the USA, such as a reduction in the capital gains tax, which also had an effect on VC). Notwithstanding these differences in terms of degree of success, the importance of start-ups, VCs, their coevolution and links with NASDAQ, Israel seemed to have followed quite closely the previously tested Silicon Valley model. Moreover, in Israel (Teubal and Avnimelech, 2002) and in other high-tech clusters (Bresnahan *et al.*, 2002), existing industry also provided the start-up segment with entrepreneurs and with significant management spillovers.

In contrast to its similarity with Silicon Valley, the Israeli pattern of VC emergence and high-tech cluster reconfiguration seems to be very different from that experienced in other recent high-tech clusters. The Cambridge (UK) cluster, for example, was probably less successful than Israel's (Breshnahan *et al.*, 2002). This may be related to significant differences in policy (both innovation support in general and specific support of VC), in the 'intensity' of start-ups and VC, in the extent of their coevolution, and in the strength of its links with US product and capital (NASDAQ) markets and with US VCs. This makes for a very significant difference, as also with India's software industry and cluster in Bangalore: while the degree of success is at least comparable to that of Israel, there are important differences. Among these we can mention emergence of a new software industry rather than 'reconfiguration' of a previously extant IT high-tech industry, as was the case in Israel; a focus on 'services' rather than 'products' and R&D; a process led by large companies rather than by start-ups and VC; and only a few linkages with NASDAQ.[9]

The studies suggest that VC should be regarded (at least as regards 'early phase' investments) as a domestic, relatively non-traded 'service' which might be stimulated or triggered once a high-tech sector exists and has attained a certain size.[10] Also the Israeli case suggests the importance of a mix of policies (Teubal and Andersen, 2000), for example a horizontal policy implemented first, which helps create favorable background conditions in terms of innovation capabilities and emergence of high technology within the business sector, and a subsequent selective policy aimed at the VC industry. Policy (more specifically, innovation and technology (ITP)) – business sector coevolution lies at the root of this process since the initial program's impact on the business sector helped identify the economy's comparative advantage in innovation and high technology and, indirectly, the specific 'needs' or priorities which a subsequent targeted program should address.[11] This process fits the systems of innovation (SI) perspective with ITP quite well (Teubal, 2002a) with its emphasis both on 'sequences of programs' and on a 'portfolio of coordinated programs (and policies)'. Both underscore the limitations of a piecemeal analysis of the impact of a single ITP program.

Objectives of the Chapter

The chapter's three main objectives are to undertake a detailed analysis of the Yozma program in the context of the evolution of high technology in Israel and emergence of a VC industry during the period 1993–8, and to develop a conceptual framework for the analysis of Yozma and its impact.

In the next section we extend our previous work on the nature and impact of Yozma (see Avnimelech and Teubal, 2002a, 2002b, 2002c) by emphasizing the integration of selected microeconomic aspects of Israel's VC industry and issues of VC organization and professionalization. In the third sections we expand the analysis by proposing a conceptual framework for the analysis of Yozma and its impact. This framework comprises a number of elements including a discussion of whether Yozma triggered VC emergence or only accelerated it.

THE YOZMA PROGRAM (POLICY PROCESS, DESIGN AND IMPACT)

General

New national priorities emerged with the beginnings of the massive immigration from the former Soviet Union during the early 1990s. The government of Israel began searching for means to employ the thousands of engineers that came into the country. Simultaneously the military industries had laid off hundreds of engineers; and many start-up companies were created, only to fail subsequently. In fact an official report of the late 1980s mentions that 60 per cent of the technologically successful OCS-approved projects failed to raise additional capital for marketing. This suggests both a capital gap and absence of marketing capabilities (also a bias towards technology in the OCS approval process).

Simultaneously officials in the Treasury (who had good undergraduate training in economics) realized that, despite massive government support for R&D, there was still a clear 'market failure' ('system failure' in our view) which blocked the successful creation and development of start-up companies. This was also related, not only to insufficient sources of R&D follow-up finance, but also to weak management abilities.

The outcome was a gradual shift in policy objectives from promotion of R&D to enhancement of start-up formation, survival and growth. This was also a response to the new model of high technology linked to the recent 'globalization' of technology capital markets. The new context involved new opportunities for peripheral economies namely the possibility for the first time and in a systematic way for high-quality start-ups to launch IPOs in global markets. This could provide not only a relatively fast return to inventors, entrepreneurs and early investors (including angels and VCs) but also the resources and exposure to penetrate global product markets (particularly the USA: see Teubal and Avnimelech, 2002). Exploiting the new possibilities, however, required changes at both the company and the systems (SI) level. Moreover their existence (or absence) increasingly became a source of country comparative advantage (disadvantage) and company competitive advantage (disadvantage). It became increasingly clear that, under the new conditions, the traditional OCS support of R&D was not enough.

The head of OCS or 'chief scientist' at the time, Ygal Erlich, pondered about ways to make OCS support more effective. Prior to the emergence of venture capital he could not find even one real success 'similar to those we see today' (interview, January 1998). The basic problem was lack of capability to grow after the product development phase. He arrived at what could be seen as a vision and strategic perspective for Israel's high technology. This involved two elements: first, the weak links in the system were both finance and marketing/management; second, the way to overcome the deficiency was to foster venture capital.[12]

A first attempt: Inbal

The Inbal program was the first attempt at implementing a specific ITP aimed at the VC industry. It was launched in 1992 one year before the implementation of Yozma. Its central idea was to stimulate publicly traded VC funds by guaranteeing the downside of their investments. The mechanism used was a government insurance company (Inbal) that guaranteed VC funds traded in the stock market (TASE), up to 70 per cent of initial capital assets. The program imposed certain restrictions on the investments of the VC companies which it covered. Four funds were established. They and the Inbal program as a whole were not a great success. Inbal fund valuations in the stock market were low, similar to holding companies' valuations. Moreover, the funds encountered bureaucratic problems and had to go to great lengths in order to prepare regular period reports. Eventually all of them attempted to leave the program, which they finally managed to do. The funds did not succeed financially and did not grow. Today all the (former) Inbal funds are held by one holding company (Green Technology Holdings).

Inbal supported publicly traded VCs with guarantees to the downside. There was no mechanism for drawing professional agents with VC abilities into the program; it did not generate VC companies with added value capabilities (including those coming from investors) and it was exposed to 'stock market sickness' and short-term thinking. The model of VC company organization was not imitated, and the 'social impact' of the Inbal program was probably very low. Policy makers and businessmen alike learned from Inbal's weak impact: the difficulty in publicly traded VCs of having investors contribute to the operation of the fund; greater difficulty in rapidly exploiting reputation earned from early exits in order to raise new capital; limits on management decision-making flexibility and on management compensation; and, last but not least, absence of incentives for the 'upside' (an important factor in attracting professional VCs).

The Design of Yozma[13]

The program began operating in 1993. The explicit objective was to create a solid base for a competitive VC industry with critical mass, to learn from foreign limited partners and to acquire a network of international contacts. It was based on a \$100m. government-owned VC fund (with the same name) oriented to two functions: investment in private VC funds ('Yozma funds' – \$80m.); and direct investments in high-tech companies (\$20m.) through the government-owned 'Yozma Venture fund'. The basic thrust was to promote the establishment of domestic, private LP VC funds that invested in young Israeli high-tech start-ups ('early phase investments') with the support of government and with the involvement of reputable foreign financial/investment institutions (generally a foreign private equity company with or without a VC arm). Such funds must be managed by an independent, Israeli VC management company. Each 'Yozma fund' would have to engage one such foreign institution together with a well-established Israeli financial institution. This emphasizes the point that the Yozma program favored entry of professional managers or of individuals with VC abilities into the infant VC industry. For an approved fund that fulfilled these conditions, the Government would invest around 40 per cent (up to \$8m.) of the funds raised. Thus \$100m. of government funds would draw \$150m. of private sector funds (domestic and foreign).

Yozma did not simply provide risk-sharing incentives to investors, as was common in other government VC support programs (it did not provide guarantees or tax benefits; nor was it accompanied by new regulations for pension funds);[14] its main incentive was in the 'upside', that is, where VC investments were very profitable. Each Yozma fund had a call option on government shares, at cost (plus 5–7 per cent interest) and for a period of five years.

The program also ensured the realization of 'supply-side learning' through the compulsory participation of foreign financial institutions ('learning from others' – a standard mechanism of infant industry development in developing countries); through participation of the Yozma Venture Fund manager (Yigal Erlich and other OCS officers) at the board meetings of all Yozma funds (they probably acted as a node in a vast information network) and through the presumed stimulation of coinvestment among Yozma funds. Culturally speaking, the stage was set for a lot of informal advising and interaction among the managers of the funds. 'Demand-side' support was provided, not by Yozma itself, but by the 'backbone' R&D

support and technological incubators programs (see Avnimelech and Teubal, 2002a). Another major point was the pursuing of an aggressive investment policy, spearheaded by Yozma Venture Fund.

A total of ten private 'Yozma funds' were created by the Yozma program, to which should be added the government-owned Yozma Venture Fund which started operating in 1993 (it managed 20 million of the $100m.; and it was privatized in 1997). Six 'Yozma funds' were founded in 1993: Gemini, Star, Concord, Pitango, Walden and Inventec; one in 1994: JVP; two in 1995: Medica and EuroFund; and one in 1997: Vertex. The total capital raised by Yozma funds was about $250m. ($100m. of it government capital) and they invested in over 200 start-up companies. Box 5.1 below summarizes the main features of Yozma's design.

An indication of the Yozma funds' success in triggering growth of the industry is their expansion, which took the form of 'follow-up' funds not supported by the Yozma program. This contrasts with Inbal funds that in most cases did not raise additional funds after establishement. Most Yozma funds (and some other funds as well, which learned indirectly from the Yozma experience) were followed by one or more funds managed by an expanding but related core of managers. Again this contrasts with the Inbal program, where no additional Inbal-type VC companies were founded after the original core of four public VCs (a few public VCs were founded in 1999–2000, but this was a result of the 'bubble'). *The total sums managed by this group amount to about $5 billion* in early 2001. This is a large share of total VC industry capital then managed. Another measure of the success is the rapid entry of non-Yozma-related funds, something triggered by the handsome profits obtained by Yozma funds, and creation by the managers of three Yozma funds of the Israel Venture Association (IVA) in 1996. Most Yozma fund managers are dominant figures in the VC industry today.

Avnimelech and Teubal (2002b, 2002c) provide an 'explanation' for Yozma, through critical mass effects and other factors, becoming the trigger for VC industry emergence and for the onset of a cumulative process of development fed by positive feedbacks and self-replication. Over and beyond favorable background conditions already mentioned and other features of the pre-emergence period, we would like to point out here the role of three additional factors: (a) the likely prior existence of 'unsatisfied demand' for VC services, a consequence of a pre-existent pool of start-ups which included some high quality firms (Checkpoint, Memco, Galileo and ESC, among others) which made a significant direct and indirect contribution to cumulativeness and emergence; (b) overlap between the learning and cumulativeness process on the one hand and the rising NASDAQ index and expanding market for communications and Internet-related equipment

BOX 5.1 CRITICAL DIMENSIONS OF YOZMA PROGRAM DESIGN

Favored type of VC company (limited partnership, closed end fund): nine of the funds adopted this form of organization, the remaining one was a public VC fund.

A focus on early phase investments in Israeli high-tech start-up companies.

Target level of capital aimed at $200–250m. (government support $100m.); this was the 'critical mass' of effort required for VC industry 'emergence'.

A multiplicity of privately owned Israeli VC funds (ten) each one managed by a local management company and involving at least one reputable foreign financial institution (and one important domestic financial institution).

Government participation in each fund: $8m. (in most Yozma Funds this represented 40 per cent of the $20m. raised).

A $20m. government fund which invested directly in Israeli high-tech companies. This VC was called the 'Yozma Venture Fund' (which should be distinguished from the Yozma program). Its aggressive investment policy stimulated investments by Yozma funds.

Strong incentive to the 'Upside' – the possibility, within a five-year period, of purchasing government's share at about cost (all but three funds made use of this option). There was no downside 'guarantee'.

Planned 'privatization' of Yozma Venture Fund took place in 1997. This previous feature ensured that the Yozma program was a catalytic program.

The Yozma program triggered a strong process of collective learning.

and software on the other (this overlap was not so consistent in other countries, where VC–start-up coevolution began operating after 1996/7 rather than in 1992/3); and (c) Yozma's successful design. Another no less important factor was the increasing globalization of capital markets, including a new trend during 1995–2000 of global flows of US VCs and US institutions investing in VCs. Some aspects of the program design and of the process leading to it are mentioned below.

Comparing Yozma with Inbal

A comparison of Yozma and Inbal will further emphasize the crucial role of Yozma's design. The programs had the same goal; their date of initiation differed by only one year; and there was a five (or more) year overlap in implementation. We can thus isolate the role of program design in explaining their differential performance (see Box 5.2 and Box 5.3).

BOX 5.2 DESIGN ASPECTS OF YOZMA AND INBAL PROGRAMS

Yozma	**Inbal**
Promoted by the OCS and mostly structured as fund of funds with the single objective of creating a VC industry.	Promoted by the Treasury and structured as a government-owned insurance company (with same name). Dual objective: promoting Tel Aviv Stock Exchange (TASE) and creating a VC industry.
Limited partnership form of VC, the ideal form of organization according to US experience and to agency theory.[15]	Publicly traded form of VC; no value added; public market sicknesses; hard to leverage current success to fundraising, low incentives for managers, bureaucracy.
Leveraged incentives to the upside. This induced professional VC teams to get organized as 'Yozma' funds. They invested in high-risk/high-expected return start-ups.	Downside guarantees, which favor entry of non-professional VC firms focused on minimizing risks rather than materializing expected returns through selection, monitoring and added value activities.
No government intervention in the day-by-day operation of Yozma funds.	Government frequently intervened and imposed bureaucratic requirements on VCs supported.
Limited period of government incentives;[16] clear and easy way out of the program.	Unlimited period of government incentives and complex way out of the program.
VC abilities were one important criterion for selection of 'Yozma	Administrative and financial criteria figured prominently in

funds'. There was flexibility in the choice of the funds. Personal recommendation of the OCS was important.	selection of Inbal VCs (there being no assurance of existence of specific VC abilities). No OCS recommendation required.
Limited number of Yozma funds planned created an incentive to join fast. This in turn contributed to creation of critical mass in two to three years.	No explicit limit (neither time nor money) to the number of funds that could enjoy the INBAL benefit.
The program was designed and implemented by the OCS which was skilled in promoting high-tech industries. It was a consensual outcome of an interactive policy process which included the Treasury, the private sector and foreign investors.	The program was designed and implemented by the Treasury which had no specific hi-tech knowledge and which emphasized financial rather than 'real' aspects. Presumed limited interaction with relevant stakeholders; and a more limited consensus among all interested parties.
Strong incentive to collective learning, to VC cooperation and to 'learning from others' (through requirement of having a reputable foreign financial institution).	No incentive to collective learning, to learning from others or to VC cooperation (legal limitations to cooperation).

BOX 5.3 FACTORS EXPLAINING THE DIFFERENTIAL YOZMA–INBAL IMPACT

Yozma	**Inbal**
Created a critical mass of VC investment in Israel.	Did not created a critical mass of VC investment in Israel.
Most 'Yozma funds' still among the 20 leading VC companies in Israel.	None of the Inbal funds are among the 20 leading VC companies in Israel.
Investments focused on early stage high-tech start-ups.	Investments also in later stages.

Yozma funds were models for the design of many other VC companies in Israel.	Very few other public traded VC companies were established in Israel.
Brought global financial and strategic investors into Israel.	Inbal did not bring any new global financial and strategic investors into Israel.
Yozma funds were involved in creating the IVA.	Not involved.
Very high private VC performance.	Low private VC performance.
Follow-up funds and strong growth of capital.	Very few secondary issues and limited growth of capital.
Yozma Venture Fund started to invest immediately. This encouraged other VCs to invest.	No mechanism to encourage VC firms to invest immediately. This explains absence of a critical mass impact.

Yozma's Impact: Macro-level Analyses

Prior to 1990 there were only two formal VC companies in Israel, the Athena fund (founded in 1985) and Star Ventures (founded in 1989). After implementation of 'Yozma' in 1993, we observe a rapid growth of the VC industry both in terms of capital raised and in terms of number of funds active in the industry (see Tables 5.1 and 5.3). During the second half of the 1990s, the Israeli VC industry becomes a significant player with a huge influence on Israel's hi-tech industry. It was then that the first foreign VC companies began to invest *directly* in Israeli start-ups.

Table 5.1 shows that a significant increase in start-up numbers occurred during VC emergence. About 750 start-up companies were founded during the period 1993–7, and many more during the late 1990s. This reflects the impact of the Yozma program and the increased availability of VC. There are direct and indirect effects. The direct impact of Yozma is reflected in the growth of (gross) *accumulated* numbers of new VC-backed start-up companies, from 110 in 1993 to 730 in 1998. After 1995, the yearly flow of new start-ups is such that it exceeds the yearly flow of new VC-backed start-ups (thus reversing the situation of 1993 and 1994). This suggests an indirect impact of VC expansion, namely an acceleration of start-up formation. All in all, we observe a sharp rise during the 1990s in the proportions of VC-backed start-up companies. Our thesis of strong VC–start-up coevolution (Avnimelech and Teubal, 2002b) is consistent with these data.

Table 5.1 Capital raised and new start-ups backed by Israeli VC firms

	1991	1992	1993	1994	1995	1996	1997	1998	1999	2000	2001	Total
Total capital raised	58	160	372	374	156	397	727	675	1752	3701	1389	9893
New VC-backed SUs	10	20	80	90	80	200	219	252	338	513	159	1961
New SUs	40	40	50	50	100	200	350	350	550	850	350	2930

Source: IVA (statistics and estimates).

Table 5.2a *Number of IPOs of Israeli companies in US and EU capital markets*

	All public offerings			VC backed public offerings		
	Number of offerings (EU)	Capital raised ($m.)	Number of IPOs (EU)	Number of offerings (EU)	Capital raised ($m.)	Number of IPOs (EU)
Before 93	~25 (0)	~1000	~22	4 (0)	~60	3
1993	18 (0)	529	16	7 (0)	103	6
1994	10 (0)	336	8	5 (0)	35	4
1995	17 (1)	614	13 (1)	8 (1)	216	5
1996	36 (5)	1025	28 (5)	17 (1)	637	12
1997	25 (3)	771	16 (3)	10 (1)	308	5
1998	17 (5)	627	14 (4)	8 (2)	215	5
1999	34 (12)	3604	20 (10)	20 (6)	1279	14
2000	35 (7)	2674	31 (6)	27 (7)	1891	24
2001	4 (1)	372	2 (0)	3 (1)	201	2
Total 90s	~220 (34)	~11550	~170 (30)	100 (19)	4885	80

Sources: Website of NASDAQ, NASE, EU capital markets, Yahoo finance and *Globes Newspaper*.

Table 5.2b Israeli high-tech companies which were targets in M&A deals, 1994–2001

	Israeli high-tech M&A deals	Average valuation	VC-backed M&A deals	Average value ($m.)
1994	2	31	2	31
1995	5	107	5	107
1996	8	86	3	39
1997	11	161	5	280
1998	23	128	7	49
1999	17	210	11	229
2000	28	361	18	503
2001	12	352	7	476
Total	106	223	58	294

Source: KPMG Israel (1997–2001) and collection of data from IVA, newspapers and other resources.

Growth in IPOs, mergers and acquisitions

Elscint's 1972 IPO (the first IPO in the USA of an Israeli high-tech company) did not signal the beginning of a new era as far as links with the US capital market is concerned. Only small numbers of Israeli companies undertook an IPO in NASDAQ (or in other markets) until the 1980s. The situation changed dramatically during the 1990s. By 2000, over 150 Israeli (or Israeli-related) high-tech companies traded in foreign capital markets (see Table 5.3), the overwhelming majority in NASDAQ. 'All public offerings' (Table 5.2a) show two significant increases in numbers: during the decade of the 1990s compared to the previous decades; and after 1995 compared to 1991–4. A final very important point is the increase in the share of VC-backed issues from roughly 30 per cent at or before 1997 to over 70 per cent in 1999–2000. The picture, which emerges, is one of increasing maturity of Israel's high-tech industry on the one hand (due to learning and other cluster effects such as the creation of the VC industry itself); and increases in the NASDAQ index on the other. Table 5.2b presents data on M&A involving Isaeli high-tech companies.

Yozma's Impact: Microeconomic Insights

Previous work on VC emergence in Israel, VC–start-up coevolution and on the role played by Yozma was largely based on industry-level data such as numbers of start-up companies, of VC companies, capital raised and invested per annum. In what follows we report on additional policy-relevant insights derived from an in-depth interviewing of 19 VC companies[17] (about

Table 5.3 Israel's high-tech cluster of the 1990s: some comparable data

	2000	1990	1980
Accumulated number of SUs created during preceding decade	~3000	~300	~150
Number of VC companies	~100	2	0
Funds raised by VCs ($m.)	3400	~49	0
Accumulated funds raised by VCs during preceding decade ($m.)	~10000	~100	0
Capital invested by Israeli VCs ($m.)	1270	~45	0
Accumulated capital invested by Israeli VCs during preceding decade ($m.)	~7000	~50	0
Capital invested in Israeli SUs (not including OCS support) ($m.)	3092	~55	0
Accumulated capital invested in Israeli SUs during preceding decade	~10000	~55	0
Accumulated No of IPOs (high-tech)	~150	9	1
Accumulated VC-backed IPOs	~80	3	1
Share of foreign sources in total SU funding	67%	NA	NA
Share of IT exports in total manufacturing exports	45.7%	~33%	~20%
Capital raised in US capital markets during preceding decade ($bn)	~10	NA	NA
Mergers and acquisitions (M&A) ($bn)	~10	NA	NA

Note: Frequently the figures in the box are approximations owing to gaps in data, multiples sources and non-official sources.

Source: SU numbers come from three sources: CBS, OCS and IVA.

50 VC funds) including eight Yozma VC companies which were founded during the pre-emergence and emergence periods.[18] On the methodological side the major thrust was the construction of indices of VC company private performance (P_p) and social impact (P_s). These were subsequently used (a) to further characterize VC emergence and (b) to further reinforce our analysis of the role of Yozma.

VC performance and social impact indicators

Ascertaining VC performance is not easy because of incomplete (and high-cost) information about events that occurred and because Israel's VC industry is young (effectively not more than eight or nine years old in August 2000).[19] This means that only the first funds raised by VC companies during phase 1 of the industry (1993–5) would have completed the 'exiting' process while other funds (both follow-up funds of incumbent VCs and first funds of more recent entrants) would only complete this in the future. Thus only a subset of VC companies and VC funds could be fully included in the analysis of VC performance and, even within this group, a full rate of return (ROR) calculation goes beyond the possibilities of this chapter. In fact, the use of 'qualitative' or 'mixed' indicators is inevitable at this stage. Also a measure of judgment will have to be applied for assigning VC companies to performance categories. We believe that, under these circumstances, the use of a number of different (including 'mixed') indicators enhances robustness of the results.[20]

A second problem concerns the distinction between what may be termed 'private VC performance' and 'social (to high-tech or to the economy) VC impact'. Absolute measures seem to be important both for private and for social VC performance. In an imperfect capital market and heterogeneous industry setting a VC contribution to private profits depends not only on its ROR but also on the absolute investment undertaken and on other absolute measures. This will also affect the social impact of the VC (for example, a very small fund with a very high ROR might still represent a small impact on the national economy and on the VC industry relative to a much larger fund with a lower ROR). Among the reasons for absolute measures of VC activity to affect social impacts are: (a) reputation effects, which might be stronger when accumulated by larger entities (Gompers and Lerner, 1999); (b) networking: most global investors would not invest in a small VC even if it had a high ROR, owing to high monitoring and transactions costs; (c) contribution to the critical mass of the industry, argued by Avnimelech and Teubal (2002c) as being an important factor in the cumulative processes leading to 'industry emergence'. Absolute output measures are also an indication of capabilities, whereas ROR could be highly affected by 'luck'. This means that a full study of the social impact of VCs should consider both direct and indirect impacts and both relative and absolute measures.

To summarize, the measures for each VC company and/or each specific fund, which we focus on here, are, first, absolute indicators: (i) number of exits relative to the date of VC foundation, (ii) IPO/M&A ratio, a measure of structure of exits, (iii) total VC capital raised; second relative indicators: (i) success ratio (number of exits divided by number of portfolio companies or investments), (ii) success ratio modified according to date of VC foundation; third, indicators of indirect effects and spillovers or externalities: (i) total capital under management (critical mass effects), (ii) best exits (reputation), (iii) whether or not reputable investors and/or strategic partners invested in the VC company, (iv) VC pioneering (introducing variety into the system) and (v) demonstration effects. The above indicators will be generated for VC companies and then averaged to obtain an overall index of private VC performance (P_p) and an overall index of VC social impact

(P_s). For 13 VC companies these are classified in one of the following categories or levels of strength: very successful (VS), moderately successful plus (MSP), moderately successful (MS) and less successful (LS).

P_p levels and P_p–P_s correlation

We now focus on the pattern of P_p and P_s during the pre-emergence (1989–92) and early emergence (1993–5) periods of Israel's VC industry. Three VCs in our sample were founded during the pre-emergence period and six during the early emergence period.

All three VCs during pre-emergence have 'high' P_p indices; and two also have 'high' P_s. Thus the performance/impact of VCs founded during this period is consistent (a) with the existence of advantages of early entry into the VC industry and (b) with the view that early entrants to a successful new industry blaze the trail for subsequent entrants. The important point here is that companies showing high private performance also generated strong social impacts benefiting VC and high-technology in the future. A rather similar, though not so overwhelming, picture exists for those VC which were founded during the early emergence phase (1993–5). Out of six such companies, four show both 'high' private performance and 'high social impact.

The high P_p and P_s indices and strong correlation among them is particularly surprising during pre-emergence. As mentioned, while high P_s might be the expected outcome of early VC activity, for example to generate a 'dynamic' stimulating subsequent emergence, there still is no assurance that P_p would be positive and high in VCs founded at such an early stage. Similarly during early emergence, the dominance of VCs with high private and social performance is not something we could have anticipated, since many firms with positive social impacts could as easily have registered weak private performance (in such a case government incentives would be responsible for positive accounting of VC profitability, and hence, for entry).

The upshot of all of this is further reinforcement of our hypothesis in Avnimelech and Teubal (2002b) concerning the importance of favorable background conditions and the pre-emergence events.

Implications for the impact of Yozma

The 'high' P_p and P_p–P_s correlation found in at least one important group of leading VC companies founded during VC pre-emergence and emergence also reinforces our view of the strong impact of Yozma. This is because most VCs with high P_s and high P_s were linked to the program and because there are ample reasons for our belief that this link positively influenced indices. Out of the eight Yozma funds in the sample, six showed 'high' P_p levels, with five of these also showing high P_s levels. Moreover, all VCs showing both high P_p and full P_p–P_s correlation (five VC companies) were associated with Yozma.

A CONCEPTUAL FRAMEWORK FOR EVALUATING YOZMA AND ITS IMPACT

In this section we consider first, defining VC emergence; second, critical events and processes which favored VC emergence or which blocked it (the 'context') and, third, whether the implementation of Yozma was critical.

Defining VC Emergence and Characterizing its 'Context'

VC emergence is a process rather than a state of affairs at a moment of time. It is characterized by (a) a high rate of entry of new VC companies and a high rate of growth of VC activity; and, in response to attainment of 'critical mass', (b) the onset of a cumulative process of growth with positive feedback effects. It converges into a state of VC industry consolidation. During emergence a lot of experimentation, and of collective and interactive learning takes place both with respect to VC strategies and procedures and with respect to VC organization. Many strategies, procedures and organizational forms do not survive; some do and are adopted by varying numbers of VCs. However, their distribution is not 'stable'. During emergence, VCs also learn and collaborate with each other no less than they compete with each other.[21] The VC industry also begins experimenting with 'institutions' and with various configurations of supporting structures. The period during which emergence takes place includes but goes beyond the PLC fluid phase (Abernathy and Utterback, 1978) when significant evolutionary variation takes place. It also includes part or all of the intermediate growth phase where evolutionary selection and reproduction operate. With VC industry consolidation, the industry attains a size which enables it to sustain a large number of supporting services. It also converges to a relatively stable distribution of strategies, procedures and aspects of organization (in Israel, with a strong focus on 'early phase' investment), a lot of which are 'embedded' in 'routines' (Nelson and Winter, 1982, ch.5).[22] Moreover, the supporting structure and set of institutions are relatively well established (IVA is an example). At this stage (which would correspond to maturity in the PLC) and for as long as external conditions remain unchanged, the VC industry, and the wider hi-tech cluster to which it belongs, effectively support the creation and growth of large numbers of new start-ups.[23]

Our study of Israel's VC industry (Avnimelech and Teubal, 2002b) shows quite conclusively that the process of emergence of the industry took place between 1993 and 1997/8.

The context of emergence

The context comprises a set of critical events and processes which operated during one or more of the three periods of Israel's VC industry identified above. These events and processes have been identified and their contribution to VC emergence (or non-emergence) assessed. Several of these contributed to two coevolutionary processes, VC–start-up coevolution (Avnimelech and Teubal, 2002b); and ITP–business coevolution (Avnimelech and Teubal, 2002d.

It is clear that there is no unique set of critical events and processes required to describe the phenomenon of VC emergence in Israel (or in other countries). In fact different authors (Florida, Kenney and Saxenian) have emphasized a somewhat different set of variables from the set we identified in our research. The main variables we utilized in our research were size, structure and achievements of the IT industry prior to and during emergence; scope of innovation capabilities accumulated prior to emergence; links and networks with world industry, product markets and capital markets prior to and during emergence; numbers of start-ups, early VC activity and VC organizational forms; strengths of universities and other non-market organizations involved in research and training; incentives programs and institutional changes (see Avnimelech and Teubal, pp.88–9). They constitute a coherent set of events and processes which is consistent with the data.[24]

Characteristics and Impact of VC-directed Policies

The VC literature has yet to deal systematically with the role of VC-directed policies in VC emergence so it should not be surprising if we do not have a clear methodology to apply to the analysis of the impact of Yozma. The detailed analysis of Yozma's design in the previous section, and the very significant quantitative share of Yozma funds (including follow-up funds), suggest that Yozma indeed had a clear impact on VC emergence. To this we may add the indirect evidence we have of a successful VC–start-up coevolutionary process, triggered by Yozma, which took place during the 1990s (Avnimelech and Teubal, 2002b). However, the caveat of the previous subsection obviously applies here: a fuller analysis requires undertaking comparative research involving both successful and failed attempts at implementing specific VC policies. The fact of VC emergence that followed implementation of a targeted VC policy is no proof that such a policy caused emergence.

For comparative research a first step would be the identification of 'contexts' which have been clearly favorable to VC emergence in certain countries/circumstances and of other contexts which have clearly not been so. A detailed analysis of these extreme cases, such as that conducted in the Israeli case, might suggest hypotheses about the role of targeted VC policies (or their absence) in the successful and failed VC cases. These would be supported by data and by appreciative theory (Nelson, 1995).

Comparative research on the impact of VC-directed policies on VC emergence should take into account 'normative' systems/evolutionary principles. The first point is the importance of comparing the policy portfolio at each phase with what could be visualized as an appropriate policy portfolio.[25] The sets of policies considered could be more or less comprehensive and this may have implications for the type of conclusion arrived at and for its precision. Only rarely will it be justified to focus on targeted VC policies one at a time without considering other VC-directed policies and 'complementary' policies (not aimed directly at VC but still having considerable influence). Complementary policies, for example, could take care of the 'demand side' for VC (as in Israel, with the technology incubators and the R&D grants programs) or, through policy experimentation and learning, could improve the policy options for the future.[26]

The second point concerns the classification of policies and programs. This should be based on a number of criteria: (a) policy objectives (for example, whether to spur VC emergence such as Yozma, or only to create suitable background conditions, as with India's policies in the late 1980s); (b) distinguishing incentives and incentives programs from institutional changes (such as deregulation of VC activity and liberalization of the stock market in India in the 1990s); (c) types of incentives (neutral or selective, supply, demand or other, to the 'downside' or the 'upside' and so on; and (d) a functional focus such as only R&D or, both R&D and other factors such as organization, management and marketing. These distinctions are crucial to analyzing a country's VC policies through time and for comparing across countries.

Further Analysis of Yozma's Impact

The microeconomic insights discussed above (pp.101–2) provide additional strength to the proposition that, despite the theoretical possibility that unaided market forces would have led to VC emergence in Israel, the Yozma program seems to have been critical for VC emergence and/or for its high economic impact. First, market forces with strong capabilities could have been necessary but not sufficient to propel high technology to its new 'Silicon Valley' configuration (including 'emergence of VC'): additional capabilities would be required and these need not automatically be generated to the extent and/or with the speed required. Second, even if Yozma only accelerated 'emergence', the economic value of the resulting high-tech transformation would have been very high, i.e., unaided market forces might have underperformed significantly compared to the Yozma-driven process. This is because of static economies of scale (which may create both low-level and high-level 'equilibrium'), dynamic economies involving externalities and spillovers flowing from learning, reputation effects, networking and other factors, and the limited window of opportunity before the collapse of NASDAQ. In all likelihood they would have created a smaller VC sector, and an associated shorter period of expansion and growth of high technology and the economy as a whole.

There are strong additional reasons for such a presumption in our case. A closer look at the idiosyncratic aspects of the VC industry will show that Yozma was necessary for emergence (rather than only enhancing its economic value). Once the key background capabilities and other factors were in place, the critical input for VC industry emergence was availability of capital, particularly intelligent capital from reputable and experienced financial institutions abroad. It has been stated during our interviews that the fact that, through Yozma, the government of Israel was willing to invest directly and indirectly in start-ups[27] was an important profitability confidence signal to such investors.[28] A second, no less important, reason relates more directly to the cumulative process generated during phase 1: a seemingly necessary condition for the first VC funds created under the auspices of Yozma to trigger entry of subsequent funds is that the former be highly profitable. Note that Yozma's design enabled Yozma funds to be highly profitable in the upside.[29] This created a strong VC reputation followed by significant expansion and new entry. Strong early profitability was due to very good exits from early investments, and this led immediately to venture capitalists worldwide and to business agents domestically to consider investing in Israeli VCs and to cooperate with them, hence the onset of cumulativeness.

CONCLUSIONS

Success in developing a venture capital industry may require adopting an evolutionary–systemic perspective not only on the processes involved before and during VC emergence but also on policy–business sector coevolution.

We suggest that the main obstacles in developing a significant high-quality VC industry are system failures (and not market failures). This implies that selective policies that aim to create a VC industry should take seriously the context in which they operate and carefully define policy objectives. A major aspect is to focus on system measures and on attracting professional venture capitalists who could guide the industry toward investment according to strict VC definitions.

Israel's successful experience took place against the background of a very favorable set of conditions, both internal and external, some of which continued to be favorable during VC industry emergence and up to 'consolidation' towards the end of the 1990s. Because of this it is our view that the Israeli experience and Israeli VC policies are not directly replicable elsewhere. What can be adopted are specific aspects of the policies implemented. Also some aspects of the evolutionary/systemic perspective used to interpret the Israeli experience may be applicable to other countries, both to 'interpret' past attempts at developing VC and as possible guides or suggestions for the future.

Very favorable conditions were being created in Israel during the 1980s and early 1990s. These included pre-existing R&D/innovation capabilities and links with US product and capital markets (Israel had a weak IPO market domestically which turned out to its advantage), achievement of macroeconomic stability and a process of capital market liberalization. Moreover, VC emergence (1993–8) was preceded by a pre-emergence period (1989–92) with significant VC-like and start-up activity. During these years important business and policy experiments/learning took place and there seems to have been a strong 'excess demand' for VC services.

Despite the prior strength of the early entrants to the industry during the pre-emergence and early emergence phases, VC emergence itself was not market-led. Rather it was triggered by a VC directed policy (Yozma). This incentive's program induced entry of high-quality, professional agents and VC management teams domestically and of significant 'intelligent' capital from abroad. This configuration and its projection (additional entry, very successful exits and so on) explain why VC emergence was very fast, why it involved strong VC–start-up coevolution and why the industry 'consolidated' or arrived at maturity after six or seven years.

Yozma was critical due to two sets of factors: first, as a means of overcoming coordination and other failures which stood in the way to achieve critical mass (collective learning, cluster effects and economies of scale) and second, as a mechanism to deal with specific VC industry characteristics and constraints flowing from the globalization process of the 1990s. More specifically, 'intelligent' and networked capital will flow to VCs operating in areas with strong high-tech 'potential' and showing outstanding returns in a short period of time. Moreover, when such a reputation effect embraces several VC companies simultaneously, it coalesces into a 'VC industry and high technology' reputation for the country.

The objective was to create an efficient VC industry which drew high-quality professionals into specialized organizations (not simply to attract venture capital). Issues of taxation were sorted out in the early 1990s; and a critical choice of supporting the limited partnership form of organization was made. Moreover, the strong emphasis on incentives to the upside was very important for attracting 'professional teams' to the emerging VC funds and VC companies.

The final design underscores the singular policy process underlying the program. First, policy makers identified a missing component of the high-tech cluster: venture capital. This was de facto characterized as a system failure. As a result of this and of strong interaction with the business sector during pre-emergence, Yozma's design made explicit a large number of parameters which seem to have been left implicit in the VC policies of other countries.

Yozma was implemented at the right time, the outcome of an evolutionary process and luck. We would emphasize two aspects: first, the overlapping of critical mass and collective learning on the one hand, with the expanding global technology product and capital markets on the other; second, implementation only after sufficient capabilities were generated (for example, after suitable experimentation and learning, and after the appearance of an excess demand for VC services).

NOTES

1. Thus the impact of VC is extremely context-sensitive since it may have a significant high-tech growth impact in established high-tech regions and a very low impact in other regions.
2. The emergent properties of the reconfigured high-tech cluster were a VC market/sector comprising large numbers of start-ups and VC, an increasing weight of start-up 'output' in total high-tech output and strong links with global capital markets
3. Avnimelech (2002) explains the role of path dependency in the enhancement of VC reputation and capabilities.
4. High tech companies existed for 20 years but a high tech industry existed for 10 years at most towards the end of the background conditions period.
5. An unintended effect of the Horizontal R&D support program was to generate awareness of the weak links in the system.
6. Our presumption is that Israeli companies through this program gained access to US product markets; and through positive feedback effects in the form of networks, reputation and links enhanced their access to US capital markets. Between 1984 and 1988 Israeli technology companies raised $300m. in NASDAQ while only $500m. was raised by all Israeli companies in the Tel Aviv Stock Exchange. We might say that a path and link to NASDAQ – so important in the 1990s – was blazed during the 1980s.
7. Self-reinforcement through positive feedback effects also resulted from early successes which, through enhanced reputation effects, led to new successes (Schertler, 2002). An

example of self-reinforcing effects concerns the indirect effects of the Yozma program on collective learning. Thus, after three years of program implementation (in 1996), the Israel Venture Capital Association was created by managers of Yozma Funds (its first director was Yigal Erlich, former chief scientist and architect of Yozma). This industry association performed some of the roles that WEMA performed in Silicon Valley during the 1970s, such as gathering and diffusion of information (Saxenian, 1998: 47–8), and the sponsoring of a systematically successful yearly meeting with individuals from high-tech start-ups, VC and other financial institutions from Israel and abroad.

8. This program extended grants to R&D performed by business enterprises from the early 1970s. These grants covered approximately 50 per cent of 'approved' costs of projects submitted to and accepted by the Office of the Chief Scientist (Ministry of Industry and Trade). Support was consistent through time and continued throughout the 1990s (with modifications) and also during VC industry emergence. It was open to all firms, independent of sector or technology.
9. We can expect a stronger similarity is the future (see Avnimelech and Teubal, 2002d) given the likelihood that a new phase in and a new segment of the Indian IT/software might emerge with features similar to those of Silicon Valley and Israel: returning nationals, product software and hardware, large numbers of start-ups and VC, and strong links with global capital markets.
10. Both of these points suggest the importance of the timing of policies within an overall evolutionary framework.
11. Within the coevolutionary framework mentioned, the former program would generate 'variation' (to a large extent 'random' variation; see Nelson, 1995) and also pave the way for identification and selection of areas where further support is required.
12. At the time, there were only two or three privately held, very small venture capital companies operating in Israel. For this (and other) reasons it was clear that the total capital available for supporting start-up activity was inadequate.
13. Most of the material below was obtained from two interviews (January 1998 and May 2000) with Ygal Erlich, the CEO of Yozma and the (or one of the most important) architect(s) of the program. Additional material was obtained from a lecture he gave at the University of Pavia in February 2001 and from other sources.
14. Capital gains tax was relatively low at the time and pension funds were allowed to invest a small amount on VC subject to government regulation. In both respects Israel's situation was a 'level playing field' with that of other countries at the time.
15. General partners have full investment and management control, a fact which provides wide flexibility in operations. LPs also have tax benefits, legal defense of investors and a direct link between owner–manager compensation and VC performance.
16. Yozma Funds could purchase governments' 40 per cent share during a period of five years after foundation.
17. In 13 of the 19 interviewed we completed assembly of all information and built reliable indices (see below).
18. Avnimelech and Teubal (2002a, 2002b) distinguish three phases of the VC industry during the 1990s: phase 1, 1993–5; phase 2, 1996–8; phase 3, 1999–2000. The VC industry emergence process comprises phases 1 and 2, while VC consolidation is phase 3. In what follows, phase 1 will be termed 'early emergence'.
19. End of the period to which our information relates.
20. Our difficulties in assessing VC 'private' performance reflect those raised in the literature (Gompers and Lerner, 1997, 1999; Mason and Harrison, 2002; Murray, 1999).
21. This is a feature of young markets. VC cooperation can take various forms, such as 'referrals', syndication. In Israel part of the informal cooperation (and maybe some of the formal one as well) takes place under the auspices or activities of Yozma.
22. Including 'change routines'.
23. According to Saxenian (1998), this is an important feature of successful hi-tech clusters.
24. It is clear to us, however, that only *comparative research* will eventually be able to confirm in a more substantial way the relevance of the above set of variables in explaining Israel's VC emergence.

25. Owing to fundamental uncertainty, lack of data and incomplete information about the basic contours of 'reality', it does not make sense here to talk of an 'optimum' policy portfolio. See Teubal (2002b).
26. Comparing VC policy portfolios across phases or through time (for example the Inbal program of 1992 with Yozma of 1993) would also be important to understand how policy experimentation and learning might have influenced the impact of VC-directed policies.
27. Directly since a portion of the Yozma program budget ($20m.) was earmarked for direct investment in start-ups through the Yozma Fund.
28. Lerner (1999) in his study of the US SBIR program (which supported government-related VC activities) found a similar phenomenon: start-ups backed by this program had a superior performance mostly due to the signaling effect which favored such companies. Note that direct government participation in a program must always convey a positive signal. The fact that it did in the early 1990s probably testifies to the reputation of the OCS.
29. In fact Yozma's design created additional incentives to VCs to select and groom very good start-ups over and beyond what the market or an outright subsidy (or government guarantee) would give.

BIBLIOGRAPHY

Abernathy, W.J. and J.M. Utterback (1978), 'Patterns of Industrial Innovation', *Technology Review*, June–July, 40–47.

Akerlof, G.A. (1970), 'The Market for Lemons: Qualitative Uncertainty and the Market Mechanism', *Quarterly Journal of Economics*, **84**, 488–500.

Allen, F. and G. Faulhaber (1989), 'Signaling By Underpricing in the IPO Market', *Journal of Financial Economics*, **23**, 303–23.

Arrow, K. (1962), 'Economic Welfare and the Allocation of Resources to Invention', in R. Nelson (ed.), *The Rate and Direction of Inventive Activity*, Princeton, NJ: Princeton University Press.

Avnimelech, G. (2002), 'Exploring VC Added Value: A Study of the Israeli VC Industry', Tel-Aviv University, MA degree paper.

Avnimelech, G. and M. Teubal, (2002a), 'Israel's Venture Capital Industry: Emergence, Operation and Impact', in D. Citendamar (ed.), *The Growth of Venture Capital: A Cross Cultural Analysis*, Westport, CT: Praeger.

Avnimelech, G. and M. Teubal (2002b), 'Venture Capital – Start Up Co-evolution and the Emergence and Development of Israel's New High Tech Cluster – Part 1: Macroeconomic & Industry Analysis', forthcoming *Economics of Innovation and New Technology*.

Avnimelech, G. and M. Teubal (2002c), 'Venture Capital – Start-Up Co-evolution and the Emergence and Development of Israel's New High Tech Cluster- Part 2: Implications of a Microeconomic Analysis', typescript.

Avnimelech, G. and M. Teubal (2002d), 'The Indian Software Industry from an Israeli Perspective: A Systems/Evolutionary and Policy View', forthcoming in E. Sridharan and A. D'Costa (eds), *The Context of Innovation in India*, London: Palgrave.

Avnimelech, G. and M. Teubal (2002e), 'Venture Capital Policy in Israel: A Comparative Analysis and Lessons for Other Countries', paper presented at the Globalization & High Tech Workshop, Institute of Advanced Studies, The Hebrew University, 8–10 December.

Barry, C. (1990), 'The Role of Venture Capital in the Creation of Public Companies', *Journal of Financial Economics*, **27**, 447–71.

Bell, M. (1982), 'Technical Change in Infant Industries: A Review of Empirical Evidence', World Bank, Washington, DC, Development Research Department, July.

Ber, H. (2002), 'Is Venture Capital Special?: Empirical Evidence from a Government Initiated Venture Capital Market', Working Paper Series, Samuel Neaman Institute, February.

Black, B. and R. Gilson (1998), 'Venture Capital and the Structure of Capital Markets', *Journal of Financial Economics*, **47**, 243–77

Bresnahan, T., A. Gambardella and A. Saxenian (2002), 'Old Economy Inputs for "New Economy Outcomes": Cluster Formation in the New Silicon Valleys', paper presented at the 2002 Druid Workshop, June.

CBS (2001), 'Development of Information and Communications Technologies in the last Decade', March.

Çetindamar, D. (ed.) (2003), '*The Growth of Venture Capital: A Cross Cultural Comparison*', Westport, CT: Praeger.

Chesbrough (1999), 'The Organizational Impact of Technical Change: A Comparative Theory of National Institutional Factors', *Industrial and Corporate Change*, **8**, 447–85.

Cooke, P. (2001), 'New Economy Innovation Systems: Biotechnology in Europe and the USA', *Industry and Innovation*, **3**, 267–89.

Dahlman, C., B. Ross-Larsen and L. Westphal (1987), 'Managing Technological Development: Lessons from the Newly Industrializing Countries', *World Development*, **16**(6), 759–75.

Dossani, R. and M. Kenney (2002), 'Creating an Environment for Venture Capital in India', *World Development*, **30**(2), 227–53.

Edquist, C. (1997), *Systems of Innovation: Technologies, Institutions and Organizations*, London and Washington: Pinter.

Florida, R. and M. Kenney (1988a), 'Venture Capital, High Technology and Regional Development', *Regional Studies*, **22**(1), 33–48.

Florida, R. and M. Kenney (1988b), 'Venture Capital-financing Innovation and Technological Change in the U.S.', *Research Policy*, **17**, 119–37.

Florida, R. and D.F. Smith (1994), 'Venture Capital and Industrial Competitiveness', Washington, DC report to the U.S. Department of Commerce, *Economic Development Administration*, May.

Florida, R., M. Kenney and D.F. Smith (1990), 'Venture Capital, Innovation and Economic Development', Washington, DC report to the U.S. Department of Commerce, *Economic Development Administration*, June.

Gans, J.S. and S. Stern (2000), 'When Does Funding Research by Smaller Firms Bears Fruit? Evidence from the SBIC Program', NBER working paper no. 7877.

Gelvan D. and M. Teuba (1997), 'Emergence and Development of a Venture Capital Industry in Israel: and Evolutionary and Policy Approach', paper presented at a symposium in honor of Alexander Volta, Como, Italy.

Gilson, R. (2002), 'Engineering a Venture Capital Market: Lessons from the American Experience', typescript.

Gompers, P. (1996), 'Grandstanding in the Venture Capital Industry', *Journal of Financial Economics*, **42**, 133–56.

Gompers, P. and J. Lerner (1997), 'Risk and Reward in Private Equity Investments: the Challenge of Performance Assessment', *Journal of Private Equity*, **1**, 5–12.

Gompers, P. and J. Lerner (1999), *The Venture Capital Cycle*, Cambridge, MA and London: MIT Press.

Gompers, P. and J. Lerner (2001), 'The Venture Capital Revolution', *Journal of Economic Perspectives,* **15**(2), 145–68.

Hodgson, G. (1999), *Evolution and Institutions: on Evolutionary Economics and the Evolution of Economics*, Cheltenham, UK and Northampton, MA, USA: Edward Elgar.

Hodson, G. (2002), 'Darwinism in Economics: from Analogy to Ontology', *Journal of Evolutionary Economics*, **12**, 259–81.

IVA 1997 Yearbook, a survey of venture capital and private equity in Israel, ed. Giza Group.

IVA 1998 Yearbook, a survey of venture capital and private equity in Israel, ed. Giza Group.

IVA 1999 Yearbook, a survey of venture capital and private equity in Israel, ed. Giza Group.

IVA 2000 Yearbook, a survey of venture capital and private equity in Israel, ed. THGC Giza Group.

IVA 2001 Yearbook, a survey of venture capital and private equity in Israel, ed. IVC.

Jeng, L. and P. Wells (2000), 'The Determinants of Venture Capital funding: Evidence Across Countries', *Journal of Corporate Finance*, **6**, 248–89.

Johnson, C. (1982), *MITI and the Japanese Miracle: The Growth of Industrial Policy, 1925–75*, Stanford, CA: Stanford University Press.

Kenney, M. (2001), 'Regional Clusters, Venture Capital and Entrepreneurship: What Can the Social Science Tell Us About Silicon Valley?', typescript in an OECD Workshop.

Klepper, S. (1996), 'Entry, Exit, Growth and Innovation over the Product Life Cycle', *American Economic Review*, **86**, 562–82.

Kortum, S. and J. Lerner (2000), 'Assessing the Contribution of Venture Capital to Innovation', *RAND journal of Economics*, **31**(4), 674–92.

Lall, S. and M. Teubal (1998). '"Market Stimulation" Technology Policies in Developing Countries: A Framework with Examples from East Asia', *World Development*, **26**(8), 1369–85.

Lerner, J. (1999), 'The Government as Venture Capitalist: The Long-Run Impact of the SBIR Program', *Journal of Business*, **72**(3), 285–318.

Lukomet, R. (2001), 'Venture Capital Funds and their Contribution to the Development of Israel's High Tech Industries' (Hebrew), Falk Institute for Economic Research in Israel, March.

Lundvall, B. (1992), *National Systems of Innovation: Towards a Theory of Innovation and Interactive Learning*, London: Pinter.

Mani, S. (2001), 'Working with the Market: The Israeli Experience of Promoting R&D in the Enterprise Sector and the Lessons for Developing Countries', United Nations University, Institute for New Technologies, Maastricht, The Netherlands.

Mason, C. & R. Harrison (2002), 'Is it Worth it? The Rates of Return from Informal Venture Capital Investments', *Journal of Business Venturing*, **17**, 211–36.

Megginson, W. (2001), 'Toward a Global Model of Venture Capital', draft, University of Oklahoma, December.

Megginson, W. and K.A. Weiss (1991), 'Venture Capitalist Certification in Initial Public Offerings', *Journal of Finance*, **46**, 879–93.

Metcalfe, S. (1996), 'Science Policy and Technology Policy in Competitive

Economy', PREST and the School of Economic Studies, University of Manchester.

Murray, G. (1999), 'Early-stage Venture Seed Capital Funds, Scale Economies and Public Support', *Venture Capital: An International Journal of Entrepreneurial Finance*, **1**(4), 351–84.

Myers, S.C. and N.S. Majluf (1984), 'Corporate Financing and Investment Decisions When Firms Have Information That Investors Do Not Have', *Journal of Financial Economics*, **13**, 187–221.

Nelson, R. (ed.) (1993), *National Systems of Innovation*, Oxford: Oxford University Press.

Nelson, R. (1994), 'The Co-evolution of Technology, Industrial Structure and Supporting Institutions', *Industrial and Corporate Change*, **3**(1), 47–64.

Nelson, R. (1995), 'Recent Evolutionary Theorizing about Economic Change', *Journal of Economic Literature*, **33**, 48–90.

Nelson, R. and S.G. Winter (1982), *An Evolutionary Theory of Economic Change*, Cambridge, MA London: Harvard University Press.

OECD (1996), 'Venture Capital and Innovation', OCDE/GD(96)168, Paris.

OECD (1997), 'Government Venture Capital for Technology Based Firms', OCDE/GD(97)201, Paris.

OECD (2000), 'The Internationalization of Venture Capital Activity in OECD Countries: Implications for Measurement and Policy', DSTI/DOC(2000), December.

Porter, M.E. (1990), 'The Competitive Advantage of Nations', *Harvard Business Review*, March–April, 73–90.

Poterba, J.M. (1989), 'Venture Capital and Capital Gain Taxation', NBER Working Paper no. 2832.

Saviotti, P. (1997), 'Innovation Systems and Evolutionary Theories', C. Edquist (ed.), *Systems of Innovation: Technologies, Institutions and Organizations*, London and Washington: Printer.

Saxenian, A. (1998), *Regional Development: Silicon Valley and Route 128*, Cambridge, MA: Harvard University Press.

Schertler, A. (2002), 'Path Dependencies in Venture Capital Markets', working paper 1120, Kiel Institute for World Economics.

Stiglitz, J.E. and A. Weiss (1981), 'Credit Rationing in Markets with Incomplete Information', *American Economic Review*, **71**, 393–409.

Teubal, M. (1993), 'The Innovation System of Israel: Description, Performance and Outstanding Issues', in R. Nelson (ed.), *National Systems of Innovation*, Oxford: Oxford University Press.

Teubal, M. (1997), 'A Catalytic and Evolutionary Perspective to Horizontal Technology Policy', *Research Policy*, **25**, 1161–88.

Teubal, M. (1999), Towards an R&D Strategy for Israel, *The Economic Quarterly* (Hebrew), **46**(2), 359–83. (English version in http://atar.mscc.huji.ac.il/-economics/facultye/teubal/rd.pdf.)

Teubal, M. (2002a), 'What is the Systems of Innovation (SI) Perspective to Innovation and Technology Policy (ITP) and how can we apply it to developing and industrialized Economies?', *Journal of Evolutionary Economics*, **12**, 233–57.

Teubal, M. (2002b), 'Observations on the Indian Software Industry from an Israeli Perspective: A Microeconomic and Policy Analysis', *Science, Technology and Society*, **7**(1), 151–86.

Teubal, M. and E. Andersen (2000), 'Enterprise Restructuring and Embeddedness:

A Policy and Systems Perspective', *Industrial and Corporate Change*, **9**(1), 87–111.

Teubal, M. and G. Avnimelech (2002), 'Foreign Acquisitions and R&D Leverage in High Tech Industries of Peripheral Economies: Lessons & Policy Issues from the Israeli Experience', *International Journal of Technology Management*, **26**(2/3/4), 362–85.

Teubal, M., G. Avnimelech and A. Gayego (2002), 'Company Growth, Acquisitions and Access to Complementary Assets in Israel's Data Security Sector', *European Planning Studies*, **10**(8), 933–53.

6. Institutional support for investment in new technologies: the role of venture capital institutions in developing countries

Sunil Mani and Anthony Bartzokas

INTRODUCTION

Developing countries have over time emerged as leading producers and exporters of high-technology products. The share of developing countries in the total world exports of high-technology products has increased from about 8 per cent in 1988 to a little over 21 per cent in 1998 (Mani, 2000). However there is considerable concentration of this activity in a few developing countries from the Asian region. In fact about 95 per cent of the developing country exports of high-technology products are concentrated in just five developing countries: Singapore, Malaysia, Philippines, Thailand and Korea. During the same period, one also sees a significant increase in the innovative activity of these countries: the number of US patents granted to innovators from developing countries increased from about 1 per cent (of the world total) to about 6 per cent (Mani, 2002). The better performance, relatively speaking, of these countries is very often attributed to the particular kind of economic policy followed by their respective governments. This policy is usually charecterized by trading regimes highly open to both foreign trade and capital. But this line of reasoning does not pay any attention at all to the considerable efforts of these countries towards strengthening the ability of their domestic enterprises to enhance their technology generating efforts.

The countries have put in place a number of institutional arrangements for this activity to flourish. While there are significant variations in the specific components of this policy across the various countries, there is one common thread that unites them, namely the use of elements of science, technology and industrial policy that explicitly aim at promoting the development, spread and efficient use of new products, services and processes in markets or inside private and public organizations (Bartzokas and Teubal,

2002). In this context, the purpose of the present chapter is to examine the role of one such institutional support mechanism for growing technology-based small and medium enterprises (SMEs). The chapter is divided into four sections. The first section undertakes a quick survey of the literature on financing new technologies. This literature has largely developed in the context of developed countries. The second section identifies one such financing mechanism, namely the venture capital (VC) institution. The conceptual underpinning of this institution and its growth across both the developed and the developing world are analysed in this section. The third section maps the structure of the VC industry in developing countries in Asia. Four different dimensions of the growth of the sector in the continent are discussed. The fourth and final section summarizes the main findings of the study.

FINANCING OF DOMESTIC TECHNOLOGY GENERATION

Technological change is the aggregate outcome of investment decisions at the firm level. In the case of SMEs, this process is driven by technological change external to individual firms. By focusing on investment decisions at the firm level we can identify significant barriers to the introduction of technological change at that level. These barriers involve credit constraints and knowledge gaps. The decision to invest in new technologies is constrained by uncertainty and information costs. Uncertainty is particularly high when technologies are new and still changing rapidly and investments are considerable. If certain categories of firms do not qualify for credit, they are more subject to exogenous shocks than if they did. Large established firms may survive thanks to better access to credit even though their profitability has eroded. But because barriers to credit stifle the emergence and growth of new firms, the new investment opportunities opened by technological change and macroeconomic adjustment are not fully taken advantage of. This effect is particularly noticeable in manufacturing exports. In addition, growth and development imply structural transformation and the emergence of new firms undertaking new economic activities. Adjustment to macroeconomic shocks similarly requires that certain economic activities and firms disappear and that others emerge in their place. If start-ups have no or little access to credit, the adjustment and growth processes are slowed. The magnitude of the new challenges and associated opportunities facing developing countries presumably helps to explain both the concentration of economic progress in some regions of the world economy and the increasing disparities in income per head across nations and regions (Bartzokas, 2001).

SMEs now account for a growing proportion of the manufacturing sector of developing countries. Especially during the last decade, the fastest growing segments within the SMEs were those based on new technologies such as information technology (IT) and biotechnology. The term 'new technology-based firms' (NTBFs) was defined for the first time by Arthur D. Little as 'independently owned businesses established for not more than 25 years and based on the exploitation of an invention or technical innovation which implies substantial technological risks'. These firms are generally located in industries such as communications, IT, computing, biotechnology, electronics and medical/life sciences. A general feeling is that these technology-based ventures, whether in the developed or developing country contexts, face extreme difficulties from the point of view of getting their projects adequately funded by the conventional capital market, whether debt or equity. Considerable attention has been paid to this aspect in the literature.

The key characteristics of NTBFs identified in the literature (Bank of England, 2001) are that: (a) their success is linked to difficult-to-value growth potential derived from scientific knowledge and intellectual property; (b) they lack in the early stages of their life cycles tangible assets which may be used as collateral; and (c) their products have little or no track record, are largely untested in markets and are usually subject to high obsolescence rates.

Funding of domestic technology generation has attracted a small but growing literature. In Figure 6.1, this literature is broadly classified into three categories. At the outset it must be made very clear that the categories in Figure 6.1 are not necessarily mutually exclusive. There is some overlap between all three and especially the latter two. Discussion of the last category is dealt with in the next section on the concept of venture capital.

The general belief is that most firms create technologies through the formal R&D route, though there has been some disenchantment with this position for some time now. The literature in this area has therefore focused on various fiscal arrangements; especially tax incentives of various sorts for encouraging firms to commit more resources to industrial R&D. Much of the literature on this theme focuses exclusively on the US situation and the main research question analysed is the efficacy of fiscal incentives for promoting R&D. A succinct review of the various international studies can be found in Hall and Reenen (2000). This literature is almost entirely based on the experience of developed countries, though there have been some sporadic attempts at extending this to developing countries as well. Mani (2002) has made a detailed study of the various tax and research grant schemes that exist in six developing countries: Korea, Singapore, Malaysia, India, South Africa and Brazil.

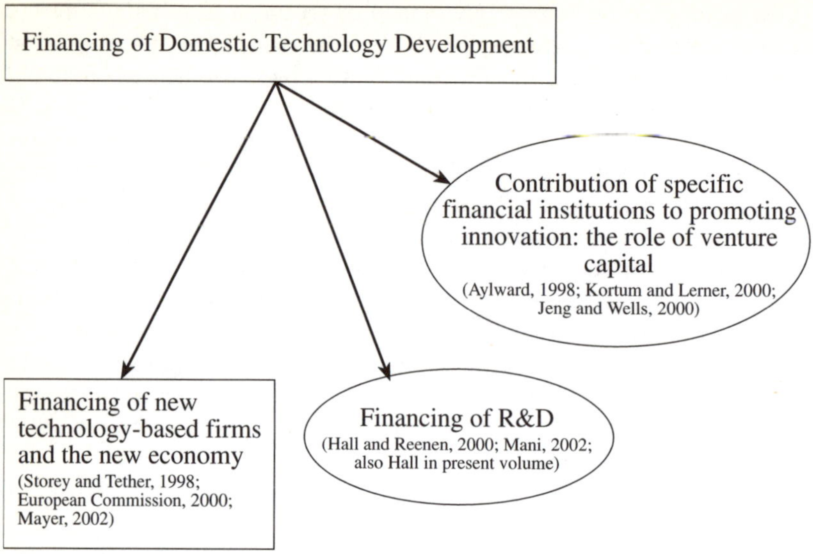

Source: Own compilation.

Figure 6.1 Classification of the literature on financing of domestic technology generation

In the recent literature on financing of the new economy, the most important work is by Mayer (2002). His paper examines the financial sector preconditions for the successful development of the high-technology sector (used synonymously with the new economy). The author examines whether the concentration of innovative activity (measured by patents) in science-based industries reflects the advantage of funding these activities through stock markets and whether the more production-oriented patenting activity in Germany relates to its highly concentrated ownership and large banking system. The main finding is that there is a close relation between types of activities undertaken in different countries and their institutional structures. Although stock markets are a very important source of development for the successful high-technology firm, they are not the most common.

Needless to add, NTBFs are vulnerable to asymmetric information about risk characteristics and default probabilities, given the fact that it is not even possible for financiers to attach probabilities to the potential outcomes of these projects. Indeed, there is a strategic complementarity between financial markets and investment in innovation at the firm level. If financial markets are underdeveloped, people will choose poorly productive, but flexible, technologies. Firms will choose technologies that are less risky, with many applications, but less productive. SMEs are reluctant to engage in sophisticated technologies as long as they cannot share the risk they incur with financial markets (Bartzokas, 2001).

The implications on financing requirements of new technology-based enterprises are evident. One of the more important papers, on incentive systems supporting new technology-based firms in the European Union, is by Storey and Tether (1998). They examine five policy instruments, one of which is direct financial support to NTBFs from national governments. The study makes a distinction between support provided in direct financial terms such as loans, grants, guarantees, tax relief and so on, and indirect support provided in the form of advisory services, access to information and so on. According to their survey only three countries, Germany, Sweden and the UK, have financial support schemes aimed exclusively and explicitly at NTBFs. These range from outright subsidies (covering up to 75 per cent of project costs) in the case of the UK to subsidized interest rates and access to funding in the case of Sweden.

A recent study by Hall (2002) has attempted to link this literature on R&D financing to the literature on venture capital and other ways of financing technology-based start-ups. The main conclusions of this study are that (a) small and innovative firms experience high costs of capital that are only partly mitigated by the presence of VC; (b) evidence of high costs of R&D capital for large firms is mixed, although these firms do prefer internal funds for financing these investments; and (c) there are limits to VC as a solution to the funding gap, especially in countries where public equity markets are not highly developed.

THE CONCEPT OF VENTURE CAPITAL

In order to overcome financial barriers to innovation, newer forms of financial institutions have been developed. VC has a number of positive features when compared to other forms of innovation financing and especially debt-financing (see Table 6.1).[1] However the most distinguishing aspect of venture financing is the rendition of a number of value added services provided by the venture capitalist to its portfolio companies. In this context, it seems particularly important to try to understand the process of venture capital investing.

VC activities as a source of risk capital for technology based ventures have been highly uneven, not only across the developed world, but even within the developing world. In fact a recent study by Kuemmerle (2001) showed that there are considerable differences in the evolution of the venture capital system in three technologically developed countries, the

Table 6.1 Distinguishing features of VC financing

	Venture capital	Debt financing
1. Objective	Maximize return	Interest payment
2. Holding period	2–5 years	Short/long-term
3. Instruments	Common shares, convertible bonds, options, warrants	Loan, factoring, leasing
4. Pricing	Price–earnings ratio	Interest spread
5. Collateral	No	Yes
6. Ownership	Yes	No
7. Control	Minority shareholders	Covenants
8. Impact on balance sheet	Reduce leverage	Increase leverage
9. Exit mechanism	Public offering, sale to third party, sale to entrepreneurs	Loan repayment
10. Value added services	Yes: (i) financial and strategic planning, (ii) recruitment of key personnel, (iii) obtaining bank and other debt financing, (iv) access to international markets and technology, (v) introduction to strategic partners and acquisition targets in the region, (vi) regional expansion of manufacturing and marketing operations, (vii) negotiating and executing mergers and acquisitions, (viii) obtaining public listing	Nil

Source: AVCJ (2000).

United States, Germany and Japan (Table 6.2). It shows that the USA alone has a well-developed venture capital system. In the USA, the pool of capital managed by venture capital firms grew dramatically during the 1980s as venture capital emerged as a truly important source of financing for small innovative firms (see Figure 6.2). According to the National Science Board (2000) VC investments in the USA have five interesting features:

1. From the mid-1990s onwards there has been a growing gap between the new capital raised and that which is actually disbursed by VC firms, implying the availability of surplus funds for investing in new and expanding enterprises.
2. Since 1990, firms producing computer software or providing computer-related services have generally received the largest share of new disbursements.

Table 6.2 Evolution of venture capital systems in the USA, Germany and Japan

	USA	Germany	Japan
Fundamentals			
Legal system	common law	civil law	civil law
Financial system	market-based/separation of commercial and investment banking	bank-based/universal banks	bank-based (with *keiretsu* ties)/separation of commercial and investment banking
Primary locus of industrial innovation	research universities, companies	research universities, companies	companies
History			
First public effort to foster enterprise creation	1958: Small Business Investment Act	Early 1960s: capital investment companies	1963: Small business investment companies
First venture capital organization involving not-for profit institutions	1946: American Research and Development Corporation (venture capital firm co-funded by MIT)	1975: WFG (semi-public venture capital firm)	1975: Centre for promotion of R&D intensive businesses (not a venture capital firm but an industry group coordinated by MITI)
First private venture capital firm	1958	1979	1973
Date of creation of first public equity market dedicated to high-growth companies	1971: NASDAQ	1997: Neuer Markt	1999: Mothers (JASDAQ, started in 1991, was not very successful)

Table 6.2 (continued)

	USA	Germany	Japan
Current state			
Number of registered private equity firms	1999: 620 (venture capital only)	1999: 172 private equity firms, of which 15 were pure venture capital firms	1999: 232 private equity firms, of which 10 were pure venture capital firms (authors' estimates)
Private equity under management	1999: $400 billion	1997: $8.3 billion	1998: $12.5 billion
of which percentage venture capital	1999: 33.6 per cent	1999: 22 per cent	1998: 16 per cent
Number of companies listed on high-growth exchange	NASDAQ: 4072 (31/1/2002)	Neuer Markt: 202 (31/12/1999)	Mothers: 10 (30/6/2000)
Market capitalization of exchange	$2873 billion (31/01/02)	NA	NA
Initial public offerings in 2001	63		

Source: Adapted from Kuemmerle (2001: 244–5), Rausch (1998).

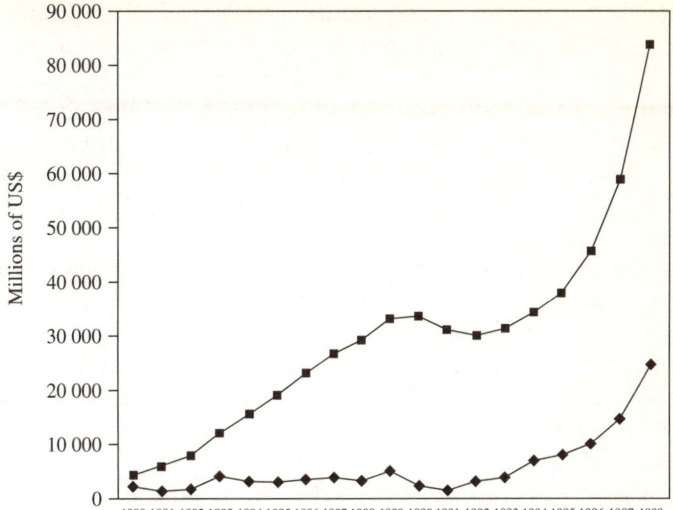

	1980	1981	1982	1983	1984	1985	1986	1987	1988	1989	1990	1991	1992	1993	1994	1995	1996	1997	1998
New capital committed	2074	1133	1546	4120	3049	3040	3613	4024	3492	5198	2550	1488	3393	4115	7339	8427	10467	15176	25293
Total venture under management	4071	5686	7759	12201	15759	19331	23371	26999	29539	33467	34001	31587	30557	31894	34841	38465	46207	59615	84180

Source: National Science Board (2000).

Figure 6.2 Trends in venture capital investments and under management in the USA, 1980–98

3. Later-stage financing (financing of expansion, acquisition or management and leveraged buy-out) accounted for very nearly three-quarters of total disbursements. Within this stage, capital for company expansion accounted for half of the total disbursement.
4. VC firms in the USA cluster around locales considered to be hotbeds of technological activity such as California, New York and Massachusetts.
5. Contrary to the popular impression, only a relatively small amount of venture capital goes to the struggling inventor or entrepreneur. Such seed financing accounted for only 5 per cent of all venture capital financing during the period 1994–8. Computer software, telecommunications, medical and health-related firms accounted for three-quarters of VC disbursements.

It is generally believed that the burgeoning venture capital industry has had a significant impact on innovation in the USA. A recent empirical study by Kortum and Lerner (2000) measure the impact VC has had on innovation and patenting in the US manufacturing sector during the period 1983–92. The authors explain that, while innovation occurs in large and small companies, projects undertaken by corporate research labs are distinct from those funded by venture capitalists. The latter scrutinize business plans thoroughly, accept only 1 per cent of them, disburse funds in stages and monitor managers extensively. The impact venture capital has on innovation is particularly strong at the early stage of financing.[2] However, in some European countries, the majority of the VC investments have gone towards consumer-related industries. So, except for the USA, VC industry is not that technology-friendly. According to Schertler (2001: 32), 'compared to the US share of investments in communications and computer-related enterprises to total venture capital, European private equity investors have only invested a small share of private equity in these high-technology enterprises'.

Recent studies have shown that there is a fundamental difference between Europe and the USA in the extent to which the venture capital industry is willing to invest in early-stage technology-based ventures (for example, Storey and Tether, 1994). Two contrasting reasons are offered to explain the comparatively low European figures. Those supplying the finance point to an absence of suitable projects, and particularly an absence of individuals with suitable managerial skills to make the project successful, as the key reason for the reluctance to invest. In contrast, those entrepreneurs seeking finance point to the technological naïveté of the financial community and the availability in Europe of comparatively high rewards for making investments in conventional sectors with which bankers are more familiar. According to the authors, there is some validity in both arguments. During the 1970s and 1980s, in both France and Sweden, there was clearly willingness on the part of financiers to invest in technology-based smaller enterprises. Unfortunately, the results were so poor that financiers subsequently became very cautious about investing in technology-based firms. This emphasizes that the selection of technology-based projects for investment is difficult. There are, of course, some specialist firms with this ability. In general and in essence, investments in technology-based firms may be deemed more uncertain, even if they are not more risky. Bankers and financiers in Europe, therefore, because they generally lack the expertise, and that expertise is expensive to acquire, have tended to favour investments outside the technology-based sector. This serves to reinforce the difficulties experienced by technology-based new and small firms in raising capital.

The growth of VC firms in the developing world is of very recent origin. In most developing countries it is not older than ten years. There are at least two reasons for this. First, most of these countries do not have one of the primary requirements for a venture capital industry, namely an exiting mechanism such as an organized market for public equity. Research (Black and Gilson, 1998, Jeng and Wells, 2000) has shown that countries which have well developed stock markets have a highly developed VC market as it provides an important form of exit to the VC investors. Second, most of them do not have large pension funds and so on, which are the main financiers of VCs worldwide. But both factors are now changing. However a major positive factor, which can pave the way for a systematic growth of the sector in developing countries, is the fact that, unlike the situation in developed countries, the main locus of innovation is individuals as opposed to companies and research institutes and universities in developed countries. This is indicated by the fact that most of the patents that are granted to developing country inventors in the USA are for individuals from these countries. Consequently the venture capital industry is now on a growth trajectory, although highly uneven as far as the developing countries are considered. In fact the industry is confined to developing countries from Asia, while its growth in Latin America and Africa (even including South Africa) is very tardy or practically non-existent. In some cases the VC industry exists merely as a subset of the private equity industry (See Table 6.3).[3]

Table 6.3 presents some interesting facts. First, the total capital under management in the US industry is almost 1.81 times the combined total of

Table 6.3 The uneven spread of VC across the world

	Total capital under management (millions of US$)	
	1991	1999
USA	NA	134400
Germany	NA	1826
Japan	15352 (115)*	21729 (221)
Israel	300 (15)	3600 (90)
Taiwan	412 (22)	4447 (132)
Korea	1547 (57)	4986 (131)
Hong Kong/China	2173 (39)	22288 (190)
India	93 (14)	1826 (44)
Singapore	868 (23)	7791 (85)
Malaysia	75 (8)	667 (28)
Indonesia	76 (8)	333 (44)
Thailand	64 (9)	265 (15)
Philippines	16 (1)	292 (14)
Australia	1231 (34)	3616 (124)
New Zealand	14 (1)**	413 (16)

Notes: * Figures in parentheses indicate the total number of VC firms; ** data refer to 1993.

Sources: AVCJ (2000); Kuemmerle (2001).

all other countries. Second, the VC industry in Germany is only as big as the one in India. Third, the VC industry in China/Hong Kong is as big as the one in Japan. Finally the Japanese VC industry has shown some significant increases and it is actually much bigger than the German one. This is an interesting result as it is generally believed that the VC industry in Japan is at best a budding one (Kuemmerle, 2001, Hurwitz, 1999).[4] Finally what is striking about Table 6.3 is the phenomenal growth of the industry in developing Asia. In the next section we analyse various dimensions of Asia's VC industry.

VC INDUSTRY IN DEVELOPING ASIA

In this section we consider the pattern of growth and specialization of the VC industry in Asia. In this study we consider the following Asian countries: Hong Kong/China, India, Indonesia, Korea, Malaysia, Myanmar, Pakistan, Philippines, Singapore, Sri Lanka, Taiwan, Thailand and Vietnam. The data are based on the twelfth survey of Asian venture capital and private equity conducted by the AVCJ (2000)[5] and the period of study is 1991–9. The dimensions of the industry considered are the following: (i) Trends in capital under management; (ii) Investment profile; (iii) Industry-wide funding; (iv) Stage of financing; (v) Source of funds; (vi) Exiting VC investments; (vii) Human resource requirement for venture capital. Finally, based on these dimensions, we develop an index of venture capital development. The index allows one to make spatial and inter-temporal comparisons of the extent to which the VC industry in a country is a solution to the financial barrier to innovation.

Trends in Capital under Management (TCUM)

At the outset the term 'capital under management' is defined as

$$TCUM = TFAI + TIPCH, \tag{6.1}$$

where

TCUM = total capital under management
TFAI = total funds available for investment
TIPCH = total investment portfolio currently held, that is cumulative total of existing investments less any divestments made.

Notwithstanding the low base, the TCUM has registered an impressive rate of growth of 26 per cent per annum (Table 6.4). Seven countries have registered above-average growth rates, but among the countries with

Table 6.4 Trends in TCUM, 1991–2001 (in millions of US$)

	1991	1992	1993	1994	1995	1996	1997	1998	1999	2000	2001	Growth rate (%)
Hong Kong/China	2173	2656	3095	6037	8044	8729	10670	15442	22288	29329	32063	30.89
Singapore	868	896	1013	1833	3164	3981	1168	5258	7791	9286	9754	27.37
Korea	1647	1629	1687	1902	2567	3224	1857	2995	4986	6020	6251	14.99
Taiwan	412	470	508	562	696	1336	1913	3598	4447	5852	6261	31.27
India	93	113	149	243	281	784	1016	1053	1826	2891	2442	38.65
Malaysia	75	147	160	194	437	448	406	460	667	587	811	26.88
Indonesia	76	57	99	225	245	289	426	328	333	169	153	7.25
Vietnam	10	22	131	247	303	276	292	258	318	157	114	27.55
Philippines	16	26	58	85	123	166	169	224	292	383	291	33.65
Thailand	64	90	98	117	165	201	177	242	265	597	580	24.66
Sri Lanka	4	16	20	49	64	67	71	68	68	46		
Pakistan	4	4	3	3	7	6	6	15	57	383		
Myanmar					112	112	113	83	36			
Average rate of growth (%)												26.32

Note: The growth rates computed here are point to point compound growth rates.

Source: AVCJ (various issues),

TCUM of at least US$1.5 billion, it is India which has registered a very high rate of growth.

There is, however, considerable geographic concentration (Table 6.5). About 95 per cent of TCUM is concentrated in just five countries, Hong Kong/China, Singapore, Korea, Taiwan and India. Within the top five, all have increased their respective shares with the sole exception of Korea, which has actually seen a significant erosion of its share, primarily because its industry has grown at a much slower rate.

Table 6.5 Geographical spread of the VC industry in Asia, 1991 and 2001 (percentage share)

	1991	2001
Hong Kong/China	40.7	54.6
Singapore	16.3	16.6
Korea	29.0	10.6
Taiwan	7.7	10.7
India	1.7	4.2
Malaysia	1.4	1.4
Indonesia	1.4	0.3
Vietnam	0.2	0.2
Philippines	0.3	0.5
Thailand	1.2	1.0
Sri Lanka	0.1	0.0
Pakistan	0.1	0.0
Myanmar	0.0	0.0

Source: AVCJ (various issues).

Investment Profile

The investment portfolio and actual investments registered sharp increases over the period 1998–9 across all countries excepting Thailand. In Thailand there has been a reduction in actual investments (Table 6.6). What is striking about the investment profile is the fact that, despite the 'Asian financial crisis', there has been no slowing down of actual VC investments in any of the countries (excepting Thailand) worst affected by the crisis. In fact the actual investments have almost doubled within such a short period.

Industry-wide funding

Following Schertler (2001), we define the following six industries as high technology-oriented industries: (i) Computer-related; (ii) Electronics; (iii)

Table 6.6 VC investments in developing Asia, 1998–9 (in millions of US$)

	Investment portfolio		Actual investments	
	1998	1999	1998	1999
Hong Kong/China	6715	8787	1378	1985
Korea	2969	3720	609	1253
Taiwan	2056	2951	881	1043
Singapore	1938	2830	424	1060
India	435	802	92	384
Malaysia	265	343	53	81
Vietnam	198	289	44	91
Thailand	234	264	74	53
Philippines	102	157	54	65
Indonesia	98	136	34	48
Sri Lanka	31	36	5	8
Myanmar	32	32	0	0
Pakistan	3	6	2	3
Total	15076	20353	3650	6074

Source: AVCJ (2000).

Information technology (IT); (iv) Manufacturing – heavy; (v) Medical and biotechnology; and (vi) Telecommunications.[6] A comparative picture of the distribution of VC investments flowing towards the high technology sector is presented in Figure 6.3. It is seen that developing Asia is marginally better than Europe on this front, but is significantly better than Japan and Australia.

Figure 6.3 hides considerable inter-country variations in the disbursement pattern (Table 6.7). Two propositions can be made with reference to technology specialization. First, among the six technology areas, many of the investments have actually gone towards computers, IT and telecommunications, while the medical industry (which includes biotechnology) has received less than 5 per cent. The only exception here is Singapore, where about 8 per cent of the disbursements have been towards this sector. Second, it is only in Taiwan and India that approximately two-thirds of the total investments have gone to the high-technology sector. Surprisingly in two other important countries, namely China/Hong Kong and Korea, only about half of the investments have been in high technology sectors.

An important hypothesis in the literature is the nexus between VC funding and the growth of certain high-technology industries such as IT (Singh *et al.*, 2000) and other high-technology sectors. The empirical evidence from both the USA and Israel substantiates this hypothesis. In order

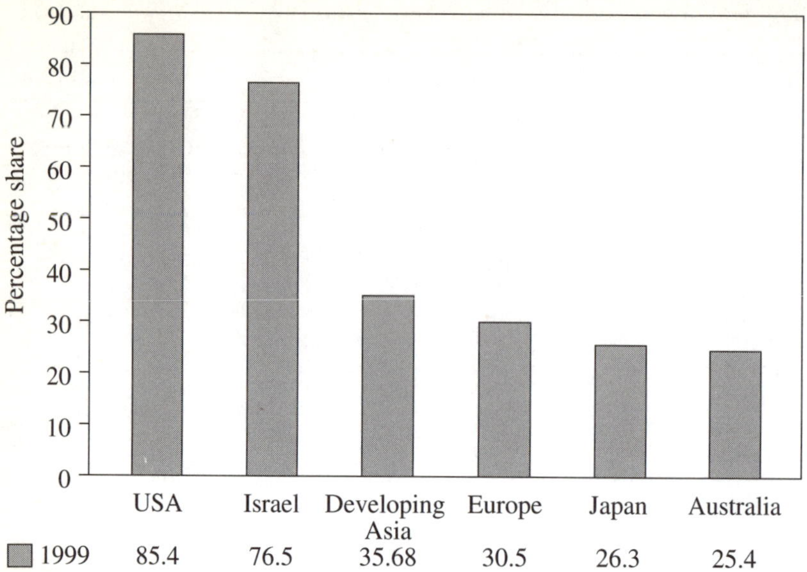

Source: AVCJ (2000), National Science Board (2000).

Figure 6.3 Distribution of VC investments flowing towards high-technology sectors

to see the link between the growth of VC investments and the growth of the high-technology sector in the Asian countries under consideration, we do two exercises, first at the macro level and second at the micro level, by taking the case of a specific country which has done excellently in terms of an index of high-technology development. At the macro level, we take into account all the Asian countries. For these countries we analyse the relationship between that portion of the VC funding that flows towards the high technology sectors and the growth of the high technology sector itself (Figure 6.4). The two variables related are the rate of growth of the VC funding towards high-technology sectors and the rate of growth of high-technology exports from these countries. Since only nine of the 13 countries in our sample are high-technology exporters, we restrict our analysis to these countries: China, India, Korea, Malaysia, Pakistan, Philippines, Singapore and Thailand. A notable omission is Taiwan. Despite this, the correlation between the two variables appears to be very high: the zero-order correlation coefficient is (+) 0.76.

It must of course be pointed out that one is not making any causation between the two. Among the countries, India has the highest correlation

Table 6.7 Industry-wide disbursement of VC investments, 2001 (percentage share)

	China	Hong Kong	India	Indonesia	Korea
Computer-related	27.4	9.0	16.8	1.2	8.3
Electronics	4.2	7.2	4.7	17.8	8.5
IT	7.7	9.9	16.7	10.5	10.3
Manufacturing – heavy	4.4	7.3	2.6	6.9	10.8
Medical/biotechnology	3.8	5.0	4.7	3.4	1.4
Telecommunications	7.7	15.3	22.6	8.1	10.7
Total	55.2	53.7	68.1	47.9	50

	Malaysia	Philippines	Singapore	Taiwan	Thailand
Computer-related	9.4	9.3	7.9	20.0	7.7
Electronics	14.1	6.0	10.1	17.8	6.3
IT	16.8	15.6	14.1	16.5	9.8
Manufacturing – heavy	10.0	1.5	2.6	2.9	8.1
Medical/biotechnology	3.5	0.0	8.8	6.3	2.6
Telecommunications	10.2	11.9	16.9	9.0	12.4
Total	64	44.3	60.4	72.5	46.9

Source: AVCJ (2001).

coefficient between the two variables (+0.81). This is highly plausible as the country has emerged as a leading exporter of computer software and, as noted earlier, about 68 per cent of the VC financing in India has gone towards the high-technology sector (Table 6.7 above). Therefore, the micro exercise is to estimate this link between VC funding (that is, high-tech VC funding) and software exports, and this is attempted below. This link between the availability of VC financing and the growth of the domestic software industry has been discussed in the literature (Miller, 2000; Baskar and Krishnaswamy, 2002). According to Baskar and Krishnaswamy (p.14), VC is 'required by most software firms in India for their sales and marketing expenses rather than for product innovation'. However this proposition needs further empirical scrutiny, as the authors have not adduced sufficient factual evidence except for a case study. The government of India and some of the state governments within the country have now established venture capital schemes specifically aimed at the software and IT industry (see Box 6.1). Given that this initiative is of very recent origin, it is too early to measure its impact on the performance of the industry. The National Taskforce on the IT industry had estimated total VC investments of

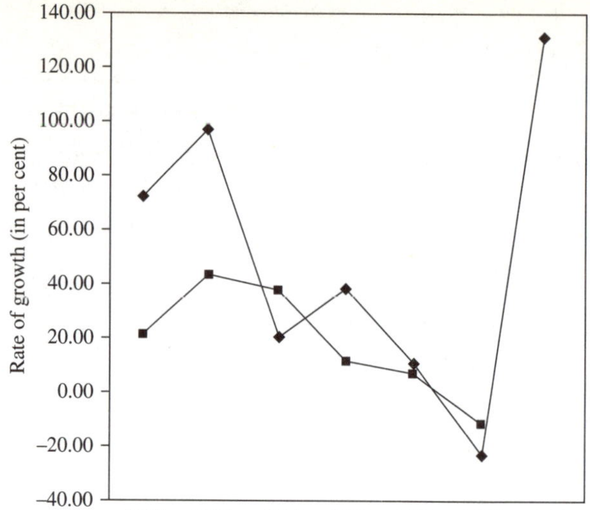

	1993	1994	1995	1996	1997	1998	1999
◆ Rate of growth of VC investments	72.38	97.28	20.39	38.46	10.79	–23.36	130.45
■ Rate of growth of high technology exports	21.3	43.5	37.8	11.5	7.0	–12.0	

Sources: AVCJ (2000), Mani (2000).

Figure 6.4 *Relationship between the rates of growth of VC funding and high-tech exports, 1993–9*

US$500 million in the five years beginning in 1998.[7] At the current rate of growth of VC funding going towards this sector, this target appears to be easily achievable (Figure 6.5). In short, VC financing is an important input for successful performance, especially in the high-technology sector.

BOX 6.1 GOVERNMENTAL EFFORTS TO ESTABLISH VC FUNDING FOR THE SOFTWARE AND IT INDUSTRY IN INDIA

Small Industries Development Bank of India (SIDBI), in association with Ministry of Information Technology, Government of India, has set up a ten-year close ended venture capital fund called the National Venture Fund for Software and IT industry (NFSIT). This was launched on 1 December 1999. NFSIT has a corpus fund of Rs1 billion (approximately US$21 million) and is a dedicated IT fund with a focus on the small-scale sector. The objective of the fund, besides meeting total financial requirements of the units, is

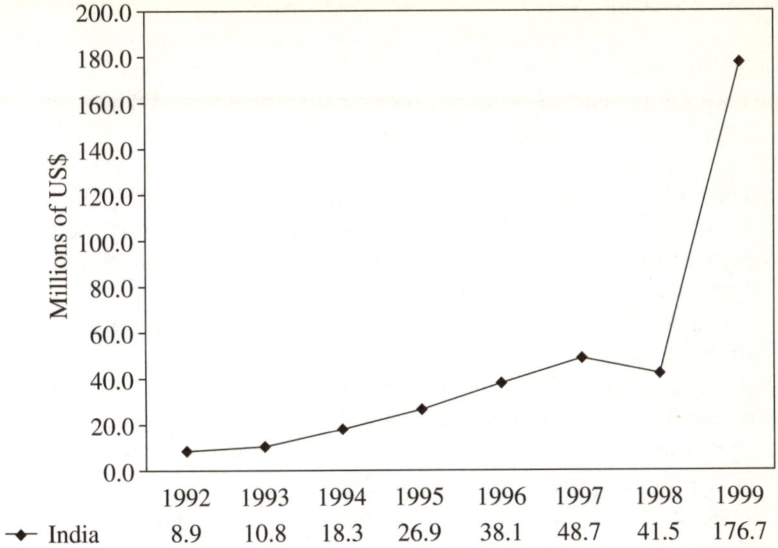

Source: AVCJ (2000).

Figure 6.5 Estimated VC financing towards the high-technology sector in India, 1992–9

to enable these units to achieve rapid growth rates and develop and maintain global competitiveness. The fund endeavours to develop international networking and enable assisted units to attract coinvestments from international venture capitalists. International linkages will help the assisted units to get a listing with foreign stock markets, such as NASDAQ, thereby achieving better valuations and offering alternative exit routes to the investors.

A portion of the fund has been earmarked for incubation projects that involve high risks and might be used for development of software products. Software products require rigorous risk evaluation for which high degrees of expertise, including international linkages, are required. The fund managed to attract a number of high-class professionals as investment managers in the asset management company.

Many state governments have already set up venture capital funds for the IT sector in partnership with local state financial institutions and SIDBI. These states include Andhra Pradesh, Karnataka, Delhi, Kerala, Gujarat and Tamil Nadu, among others.

Source: http://www.sidbiventure.co.in/svc-01r.htm.

Stage of financing

In a recent succinct review of the facts that are known about venture capital activity, Gompers and Lerner (2001) have identified four different factors that affect the financing of young firms:

1. There is a tendency for the paid managers of a firm to indulge in wasteful expenditures if the firm raises equity from outside investors. This is because the manager may benefit disproportionately from this activity and does not have to bear its entire cost.
2. Likewise, if the firm raises debt, the manager may increase risk to undesirable levels.
3. It may be difficult to write a contract governing the financing of a firm if the effort of the entrepreneurial firm cannot be ascertained with complete confidence. This arises in a situation where all the outcomes of the firm cannot be correctly predicted.
4. The above three gives rise to uncertainty and informational asymmetries in the case of young firms.

The above four factors are likely to be acutely significant for companies with intangible assets and whose performance is difficult to assess. A very good illustration of this is in the case of high-technology companies with a heavy reliance on R&D, especially in their early stages (seed, start-up and first-stage financing). The theoretical expectation is that specialized financial intermediaries such as venture capital institutions are designed to reduce these information gaps and thus allow firms to receive the financing that they cannot raise from other external sources.[8]

Mayer (in Chapter 3 of this book) has outlined the sources of capital for a young high-technology firm through its various stages of growth At the initial stages of such a firm a lion's share of the financial input emanates from savings of the entrepreneurs and of their family members and relatives. Whatever external equity these firms are able to generate is raised from informal venture capital sources, referred to as 'business angels' (wealthy or reasonably wealthy private investors). In actuality, even in the best situation, namely the US case (Figure 6.6) not more than a quarter of the VC funding has actually gone towards this stage. In fact there has been considerable erosion in the share of all early stages in total disbursements over time.

The factors that determine the different contributions of business angels and venture capitalists to start-up financing has been examined in Van Osnabrugge (2000). This was done by comparing the initial screening, due diligence, investment criteria, contracts, monitoring and exit routes

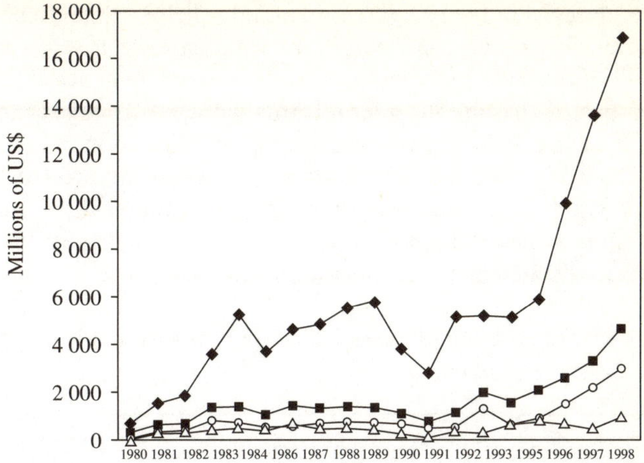

	1980	1981	1982	1983	1984	1986	1987	1988	1989	1990	1991	1992	1993	1995	1996	1997	1998
Total disbursements	703.3	1559	1902	3651	5293	4686	4888	5603	5835	3869	2875	5229	5236	5946	9897	13558	16778
Early stage	336.2	686.8	714.1	1396	1446	1491	1416	1470	1416	1148	825.8	1186	2100	2143	2658	3373	4700
Seed	11	47.7	63.1	111.4	129.7	117.6	122	144.4	184.8	124.5	88	158.2	314.2	312.5	376.8	629.3	717.1
Start-up	158.2	296.5	293.5	443.7	558.2	746.2	529.9	543.7	441.6	293.8	171.3	448.1	412.6	901.6	732.8	525	974.5
Other early-stage disbursements	166	342.5	357.5	840.9	758.5	627.4	763.9	781.6	789.8	729.3	566.5	579.6	1373	929	1549	2218	3009

Source: National Science Board (2000).

Figure 6.6 Share of early-stage financing in total VC disbursements in the USA, 1980–98

employed by the different types of investor. The results of the analysis (quoted by Mayer, Chapter 3 in this volume) showed two important differences between the two groups. First, venture capitalists act like institutions following principal–agent relations of limiting risks through monitoring, while business angels place more emphasis on ex post involvement. This differential behaviour is very much a function of the nature of ownership of VCs. Second, from the very outset VCs are focused on exit and expect a much higher rate of return than business angels: in the UK these were almost double the rates expected by business angels. It is against this background that we analyse the stage of financing of VC firms in the developing Asian context.

The Asian countries have a much better performance in terms of funding projects at their early stage (seed and start-up). In fact the weighted average ratio of early stage financing to later stage for all the countries in the sample is 0.45. Only the ratio for India is greater than unity, implying a preponderance of early-stage financing. China is the only other country which has a ratio greater than the average. The availability of good-quality high-tech projects in India and China may be an explanation. Also most of the Indian venture capital firms were initially based on governmental funding, though this has changed in the recent period and this helped them to be really venturesome. This shows that, despite its small size, relatively speaking, the Asian VC industry is showing signs of sound development. Within the early stage, much of the funding goes towards the start-up stage (Table 6.8) and within the expansion stage (which consists of expansion, mezzanine – also known as bridge finance – buy-out/buy-in, turnaround/restructuring) it is the expansion stage which accounts for the maximum share. Lack of consistent time series data does not allow us to track any intertemporal changes in disbursements across the various countries.

Table 6.8 Distribution of stage-wise disbursements of VC in Asia, 2001 (percentage share)

	Seed	Start-up	Early stage	Expansion	Ratio of early stage financing to expansion stage
China	8	33	41	59	0.695
Hong Kong	4	19	23	77	0.299
India	7	36	43	57	0.754
Indonesia	2	9	11	89	0.124
Korea	7	19	26	74	0.351
Malaysia	4	29	33	67	0.493
Philippines	1	15	16	84	0.190
Singapore	2	26	28	72	0.389
Taiwan	9	26	35	65	0.538
Thailand	3	15	18	82	0.220
Vietnam	5	26	31	69	0.449

Source: AVCJ (2001).

Source of Funds[9]

In terms of source of funds, the situation in developing Asia is somewhat more similar to Japan and Israel than to the USA (Table 6.9). Corporations account for the single largest share, while in the USA the sources are fairly equally distributed. Pension funds, which are an important source in the US case, are not in the case of Asian countries. This is because the size of pension funds in most of these countries is not sufficiently large and regulations prevent them from investing in VC funds. Government agencies form only about 10 per cent. This shows that, in most developing countries, the VC funds are privately backed. The micro picture is, of course, quite different (see Table 6.10).

It is only in Malaysia that the government is an important source of

Table 6.9 Source of funds to VC, 1999 (percentage shares)

	Developing Asia*	Australia	Japan	Israel	USA
Corporations	47	8	54	39	16
Insurance companies	17	7	17	11	11*
Banks	13	10	18	22	0
Government agencies	10	13	3	13	0
Pension funds	6	55	4	4	18
Private individuals	5	6	2	8	19
Endowments	0	0	0	0	15
Others	2	1	2	3	22***
Total	100	100	100	100	100

Notes: * Weighted average of all developing Asian countries excluding the Philippines; ** refers to both insurance and banks; *** includes foreign investors also.

Sources: AVCJ (2000); Gompers and Lerner (2001).

funds. The majority of the capital for the Asian VC companies has emanated from domestic sources (Table 6.11). But there are some notable exceptions to this, such as China, Hong Kong, India and Singapore. In these cases much of the VC has actually originated from western sources. This shows that intra-Asian investments are quite limited.

Governmental programmes of various sorts have played an important role in establishing and 'pump priming' the VC industry even in developed countries. A survey of these in the OECD countries can be found in OECD (1998) and Jeng and Wells (2000). These governmental schemes vary from providing legal infrastructure and specific tax exemptions to establishing funds that invest directly in private equity projects. Among the developing countries in our sample, there has been explicit governmental support for establishing a VC industry in all, and especially in India and Singapore. A detailed survey of this can be found in Mani (1997, 2002).

Apart from the domestic governments of these countries, a major impetus for the establishment and growth of the VC industry in developing countries has emanated from multilateral institutions such as the World Bank and one of its affiliates, the International Finance Corporation (Aylward, 1998; Pfeil, 2001). In fact, as Mani (1997: 232) documents, the genesis of the VC industry in India can be traced to a series of efforts by the World Bank in the 1980s as part of its 'Industrial Technology Development Project' in India. As part of this project a loan of $45 million was made available to the government to support four

Table 6.10 Source of funds to VCs, by country, 1999 (percentage shares)

	China	Hong Kong	India	Indonesia	Korea	Malaysia
Pension funds	7	9	2	9	3	1
Corporations	42	43	61	34	49	30
Banks	18	8	15	8	18	17
Government agencies	12	6	9	19	12	45
Insurance companies	18	30	7	13	9	5
Private individuals	1	4	5	2	4	1
Others	1	0	0	15	5	2
Total	100	100	100	100	100	100

	Myanmar	Pakistan	Singapore	Sri Lanka	Taiwan	Thailand	Vietnam
Pension funds	0	0	5	1	1	7	16
Corporations	40	37	43	33	67	37	33
Banks	23	31	14	50	7	36	31
Government agencies	15	8	19	4	1	0	3
Insurance companies	7	8	10	5	8	11	11
Private individuals	0	12	7	6	15	6	1
Others	15	4	2	1	1	3	5
Total	100	100	100	100	100	100	100

Source: AVCJ (2000).

Table 6.11 Geographical breakdown of VC sources, 1999 (percentage shares)

	Domestic		Foreign	
	1999	2001	1999	2001
China	28	53	72 (39)	47 (24)
Hong Kong	7	11	93 (77)	89 (67)
India	34	21	66 (60)	79 (60)
Indonesia	42	60	58 (40)	40 (24)
Korea	75	65	25 (20)	35 (28)
Malaysia	51	51	49 (25)	49 (26)
Myanmar	18	NA	82 (12)	NA
Pakistan	91	NA	9 (9)	NA
Philippines	89	42	11 (1)	58 (43)
Singapore	27	34	73 (38)	66
Sri Lanka	75	NA	25 (16)	NA
Taiwan	85	85	15 (11)	15 (14)
Thailand	35	20	65 (43)	80 (66)
Vietnam	14	13	86 (80)	87 (76)
Australia	89	83	11 (5)	17
Israel	78	NA	22 (19)	NA
Japan	82	87	18 (15)	13 (8)
Average for Asia	49	49	51 (39)	51 (37)

Note: Figures in parentheses indicate the share of non-Asian countries.

Source: AVCJ (various issues).

venture capital entities financing technologically innovative and growth-oriented small enterprises. The government, however, lent on this amount to four state-owned venture capital firms.

Exiting VC Investments

VC investments being primarily in the equity of the investee firm, the return to the VC is in the form of capital gains to be made while offloading the shares at a later date, when the venture has achieved some maturity. It is this capital gain to be made that brings in a return to the VC. There are at least five main exit options (British Venture Capital Association, 2002): see Figure 6.7.

Of the five, the two most important and commonly used exits are trade sales and the IPO routes. In the USA, 56 per cent of all IPOs (in 1999) were venture-backed, while in terms of money the proportion was about a third

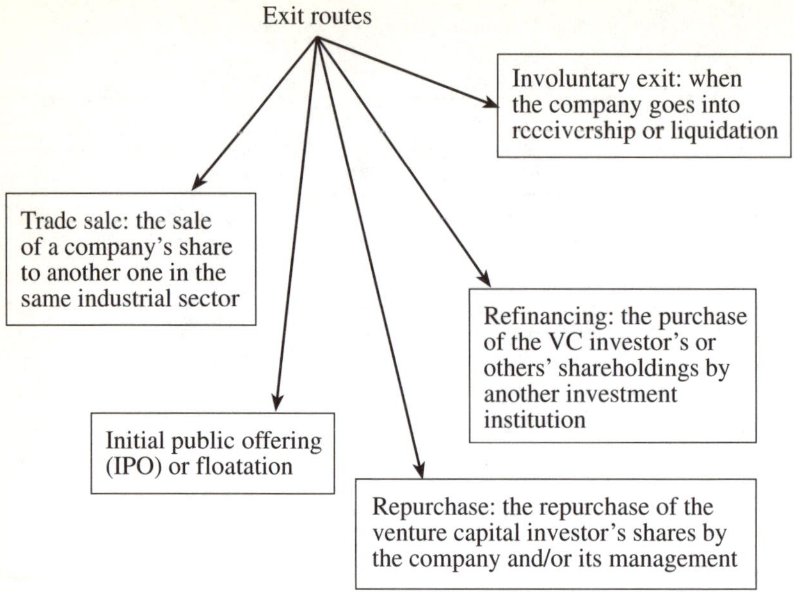

Source: British Venture Capital Association (2002).

Figure 6.7 Exiting from VC investments: the five routes

(Gompers and Lerner, 2001). According to Jeng and Wells (2000), IPOs are the most attractive option for liquidating the funds.

Korea has the highest rate of divestment (defined as the amount divested per year taken as a percentage of TCUM). This is presented in Table 6.12. However, some of the major countries, such as China, Hong Kong and India, had low rates of divestment. Given that the data only refer to four recent years, it is rather difficult to draw any firm conclusions. Whether the higher rate of divestment in the East Asian countries is due to the financial crisis requires further research. Alternatively the low rates of divestment in some of the major countries may in fact be a reflection of the regulatory framework with respect to the lock-up period[10] (Mani, 2002) and the easy availability of the exit routes. The various exit mechanisms followed in the Asian situation are summarized in Table 6.13. IPOs are the only exit route in most of the countries and therefore, as noted by Jeng and Wells (2000), this is indeed one of the determinants of VC investment.

In fact Jeng and Wells (2000) identify two specific reasons why IPOs are an important source of exit to a VC investor. First, according to the literature, the most attractive option for exit is through an IPO. A study by Venture Economics (1998) quoted in Jeng and Wells (2000) finds that

Table 6.12 Divestment rate in VCs, 1998–2001

	1998	1999	2000	2001
Korea	14.47	13.24	16.1	7.81
Indonesia	2.30	6.41	17.2	9.80
Thailand	5.13	5.36	1.7	2.76
Taiwan	3.24	4.38	2.5	3.58
Philippines	1.43	2.12	10.2	14.04
Singapore	1.13	1.89	4.1	2.52
Sri Lanka	1.71	1.77	2.2	NA
Malaysia	0.34	0.39	1.0	0.49
India	0.20	0.35	1.0	12.53
China	1.57	0.27	1.2	1.74
Hong Kong	2.73	0.02	2.9	1.84
Australia	5.50	6.51	6.20	9.08
Japan	11.89	4.03	5.47	4.01

Source: AVCJ (various issues).

US$1.00 invested in a firm that eventually goes public yields a 195 per cent return for a 4.2 year average holding period. The same investment in an acquired firm only provides an average return of 40 per cent over a 3.7 year average holding period. Second, if regaining control is important to an entrepreneur, IPOs are the best choice, given the fact that the other options such as trade sales frequently result in loss of control. The empirical work of Black and Gilson (1998) statistically established, for the first time, a direct positive link between the existence of a well developed stock market and IPOs and the growth of VC financing, though of course the study was restricted to the US case.

Building on this, Jeng and Wells established the same result for a group of 21 developed countries over the period 1986–95. Apart from IPOs, they also included six other independent variables, namely accounting standards, labour market rigidity, market capitalization and GDP growth, availability of private pension funds and government support programmes. Among all these the IPO variable turned out to be most important determinant of, especially, later-stage VC investment across the selected countries.

Our own qualitative study (Table 6.12) shows that this is indeed the likely case in our sample of developing countries. Since data on country-wide IPOs are not readily available,[11] we are constrained to limit the analysis to only one country, India. For India, we relate the rate of growth of annual VC investments during the period 1993 to 1999 to the corresponding IPOs (actual subscription of issues by new companies). The results are presented

Table 6.13 Profile of exit routes for VC firms in Asia, 1998

Country	Market capitalization (millions of US$)	Size of the IPO market (millions of US$)	Remarks
1. China	284 766 (853)*		Exits in China remain problematic; only one stock exchange; handful of Chinese companies are listed on the Hong Kong stock exchange and on NASDAQ
2. Hong Kong			IPOs and trade sales constitute the most frequently used exit routes; two stock exchanges, the second one being the Growth Enterprise Market launched in 1999
3. India	64 498 (5 860)	1 796 (1 575)**	There are 22 stock exchanges; an OTC was established in 1992; some high-tech companies are listed on NASDAQ
4. Indonesia	9 709 (287)		The Pakades (government-assisted) guidelines outline multiple divestment avenues for VC firms; exits can be made via the capital market, private placements and the sale of shares; an OTC was established in 1994
5. Korea	137 859 (748)		Trade sales and IPOs are the viable routes; OTC (Korean Securities Automated Quotation System) was established in 1996
6. Malaysia	28 889 (736)		A new OTC (MESDAQ) was established in 1999, but attracted only one listing
7. Pakistan	5 418 (773)		IPO is the only route
8. Philippines	9 992 (221)		Trade sales and IPOs are the viable routes; companies prefer regional exchanges
9. Singapore	94 469		IPOs and trade sales; an OTC (SESDAQ) was established in 1986; two Singaporean companies are listed on NASDAQ
10. Sri Lanka	281 (233)		IPOs, buy-back of shares and the sale of shares to third parties are the main routes

11. Taiwan	884698 (437)	IPOs are the main exit route; more than 200 venture-backed Taiwanese companies are listed on the US OTC market
12. Thailand	20734 (418)	The main exit route is trade sales. An OTC was established in 1995. About 400 firms were listed (as of 31 December 1999). The other exchange, the Securities Exchange of Thailand (SET) also allowed the floatation of Vietnamese companies and Thai–Vietnamese joint ventures. The main exit route, however, is trade sales and not IPOs
13. Japan	2495 (757)	
14. Australia	874 (283)	
15. Israel	39 (628)	

Notes: * Figures in parentheses indicate the number of companies that are listed on the stock exchange; ** figures in brackets indicate the size of market capitalization in millions of US$ of India's OTC exchange.

Source: AVCJ (2000); http://www.sebi.gov.in/pmd/aprmar01.htm; International Finance Corporation (1999).

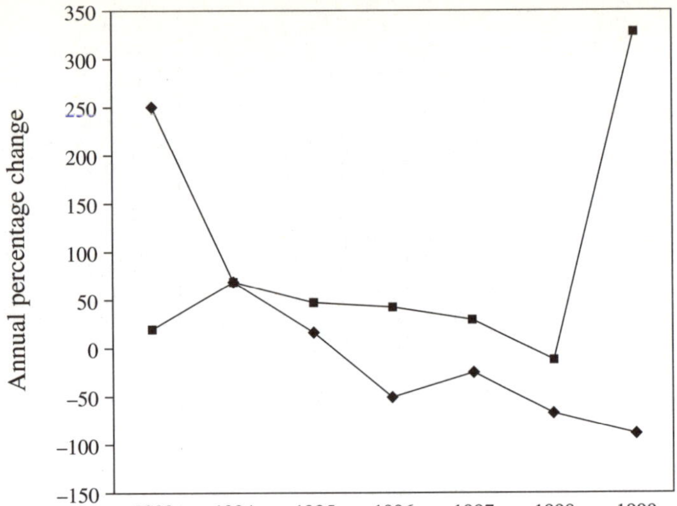

	1993	1994	1995	1996	1997	1998	1999
◆ Rate of growth of IPOs	251.83	68.54	15.85	–51.79	–27.15	–70.27	–91.38
■ Rate of growth of VC	20.71	69.80	46.94	41.61	27.75	–14.75	325.99

Sources: AVCJ (2000), Reserve Bank of India (http://rbi.org.in/sec7/25891.doc?).

Figure 6.8 Relationship between IPOs and VC investment in India, 1993–9

in Figure 6.8, which shows a rather high degree of positive correlation between the two variables, if one omits the two terminal years of 1993 and 1999. However, more sophisticated tests are required before one can draw any firm conclusions about the importance of IPOs for the successful growth of VC financing in developing countries.

Human Resource Requirement for Venture Capital

Venture capitalists spend a large block of their time screening potential investments, much like a bank loan officer evaluating a loan application. Also, like investment bankers, venture capitalists act as consultants for their portfolio firms. However, the consulting side of the venture capitalist's relationship with its portfolio firms is more important than it is for other financial intermediaries, since the start-up firm's management is often inexperienced. The venture capitalist expects to have an intense involvement with each portfolio firm for three to five years, by which time the successful firm is merged with or acquired by another firm or goes public in the secondary market. The successful venture capital expert commands a

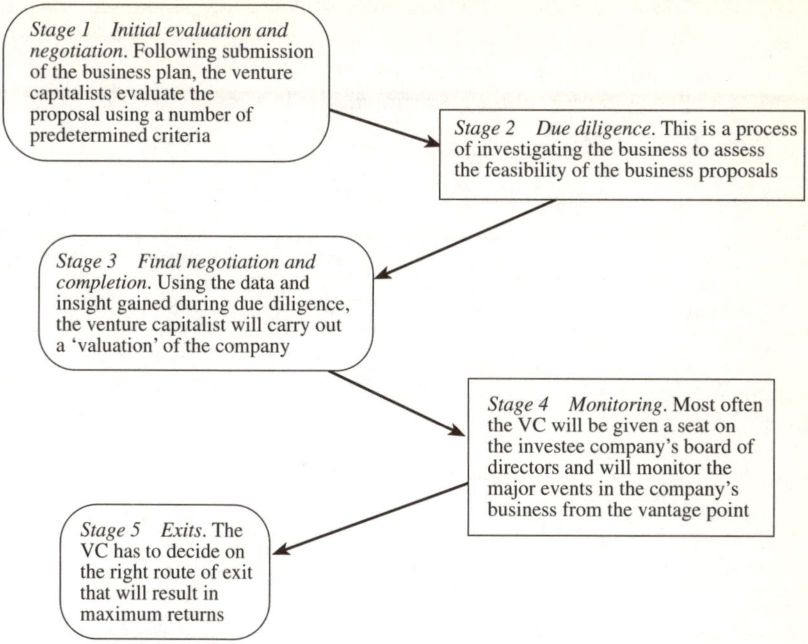

Source: AVCJ (2000).

Figure 6.9 The five stages in a typical VC investment process

unique mix of skills including securing financing, evaluating ideas and managers and building successful management teams.

VC, as we have noted, is a specialized form of financing technology-based ventures, so the human resource requirement for this sort of agency is much more than other forms of financial institutions such as normal commercial and development banks. Venture funds typically look for talented managers who can spot clever new ideas around them. It is not difficult to understand that, without these business leaders, ideas remain untapped and funds dormant. The sophisticated nature of the human resource requirement for the VC industry may easily be gleaned from the five different stages in a typical VC investment process (Figure 6.9).

Given the fact that most countries are new to this industry, lack of availability of VC professionals can be an important detriment to the successful growth of the industry. Although the total number of VC professionals has shown some increase (Table 6.14), except for Hong Kong/China, the rate of growth is not significant. In fact, when taken as a density figure, the number of VC professionals per unit of TCUM is not only lower in the developing Asian countries, but also has shown some slight reductions. Even in countries like India there have been serious shortages of VC personnel, especially at steps 2 and 3 (see Figure 6.9).[12] So this is another aspect which may require some governmental intervention to create an adequate pool of VC professionals.

Table 6.14 Venture capital professionals, 1988–2001

	1998	1999	2000	2001
Hong Kong/China	494	609	987	991
Singapore	331	453	521	509
Korea	409	435	512	483
Taiwan	356	381	428	437
India	212	247	278	267
Malaysia	59	78	99	95
Indonesia	189	178	135	105
Vietnam	28	30	23	21
Philippines	44	60	64	59
Thailand	49	71	62	58
Totals	2171	2542	3109	3025

Source: AVCJ (various issues).

Index of VC Development

The main purpose here is to develop an index of VC development in each of the major Asian countries. Such an exercise, we feel, will enable us to make a proper benchmarking of the industry in Asia. This index captures the extent to which a country's VC institutions are developed to service technology-based ventures in the early stage of their existence. This is called the index of VC development (VCDI). It consists of two separate indices combined into one. The two separate indices are the finance index (FI); and the technology index (TI). The FI captures the extent to which the VC institutions in a particular country finance early stage ventures, that is, at the seed and start-up stage. It is computed by taking the relative share of this stage in the total VC disbursements in a specific country during a particular time period (one year). The TI, on the contrary, captures the extent to which the total disbursements flow towards high-technology sectors, namely computer related, IT, medical and telecommunications sectors in a specific country during a particular time period (one year). For any component of the VCDI, the individual indices can be computed according to the general formula:

$$\text{Index} = \frac{\text{Actual } x_i \text{ value} - \text{mininum } x_i}{\text{Maximum } x_i - \text{minimum } x_i} \tag{6.2}$$

For both the FI and TI, we assume that the maximum and minimum values (in percentage terms) are 100 and 0, respectively.

The VCDI is conducted in two steps. In the first step, we construct the FI and TI for each of the countries in our sample for two years, 1999 and 2001.[13] In the case of the FI, the x_i value is the percentage share of early-stage financing in total disbursements during a year and in the case of TI is the percentage share of total financing going towards the high-technology sectors. In the second and final step

$$VCDI_C = (FIc \times w_i TIc) \tag{6.3}$$

The $VCDI_C$ thus computed for the selected countries for the years 1999 and 2001 is charted in Figure 6.10. The following inferences can be drawn: (a) the index has shown an improvement in all countries excep one, namely Korea in the latter period; (b) it appears that the VC industry in Malaysia shows considerable improvement in the latter period.

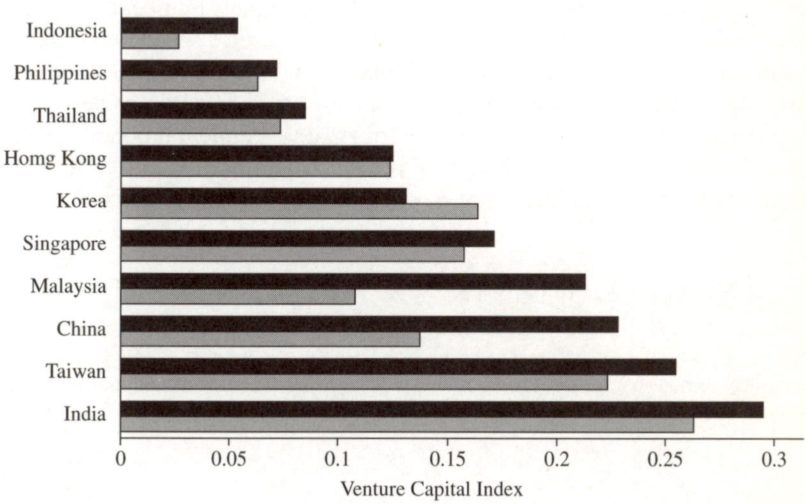

	India	Taiwan	China	Malaysia	Singapore	Korea	Hong Kong	Thailand	Philippines	Indonesia
■ 2001	0.29283	0.25375	0.22632	0.2112	0.16912	0.13	0.12351	0.08442	0.07088	0.05269
▨ 1999	0.26136	0.22156	0.13596	0.10697	0.15568	0.16211	0.12264	0.07282	0.06244	0.02645

Source: Computed from data provided in AVCJ (Various issues).

Figure 6.10 Index of VC development, 1999 and 2001

A year-wise computation of the VCDI will allow us to chart the growth trajectory of the VC industry in a specific country compared to the rest of the world, especially as a source of finance to technology-based firms in their early stage.

CONCLUSIONS

The role of VC as an input to innovation is now a more or less accepted fact, though the empirical substantiation for this statement has come from advanced countries. This chapter is an attempt to extend this line of reasoning to developing countries. Although uneven in its spread across countries, the concept of VC is now spreading fast to most countries and especially to those countries which have well developed exit mechanisms such as a reasonably well functioning stock market. An examination of the relationship between VC investments and the growth of the high-technology sector shows a positive relationship between the two. This macro exercise has been further substantiated by a micro one by taking the specific case of India, which has emerged as a successful exporter of computer software. This is a hypothesis which needs further empirical scrutiny. The successful growth of the industry also requires the availability of an adequate number of VC professionals and an avenue such as IPOs for a proper exit. The study concludes by constructing an index of VC development which allows one to benchmark the degree of VC development in a country against that of best practice. Such a comparative analysis of performance should aid policy makers in redirecting the efforts of their local venture capital institutions by making them more effective financiers of new technology-based enterprises.

ACKNOWLEDGMENTS

We thank Djono Subagjo and Ad Notten for their help in writing this chapter. However, we are solely responsible for any errors or shortcomings that may still remain and the views expressed are those of the authors and do not necessarily reflect the views of the United Nations.

NOTES

1. Venture capital is usually referred to as one type of private equity investments. According to Jeng and Wells (2000: 243), 'private equity investments are investments by institutions or wealthy individuals in both publicly quoted and privately held companies. Private equity investors are more actively involved in managing their portfolio companies than regular, passive retail investors. The main types of financing included in private equity

investing are venture capital and management and leveraged buyouts'. Except for the USA, and especially in Europe, this distinction between the two is not usually made. In the USA, VC as a percentage of total private equity increased from 18 per cent in 1993 to 43 per cent in 1999. See Pfeil (2001) for the details.

2. The authors regress a measure of the number of successful patent applications, in each industry, against a measure of the number of firms that obtained venture capital backing and against total disbursements. Patenting patterns across industries over a three-decade period suggest that the effect is positive and significant. The results are robust to different measures of venture activity, subsamples of industries and representations of the relationship between patenting, R&D and venture capital. Averaging across regressions, the authors come up with an estimate, for the impact on patenting of a dollar of venture capital relative to a dollar, of 3.1, and this estimate suggested that VC accounted for 8 per cent of industrial innovations in the decade ending in 1992. Further, according to the authors, given the rapid increase in venture funding since 1992, and assuming that the potency of venture funding has remained constant, the results imply that, by 1998, venture funding accounted for about 14 per cent of the innovative activity in the USA (see Kortum and Lerner, 2000).
3. The VC industry in South Africa is an example of this. See Mani (2002) for the details.
4. The term 'venture business' first surfaced in Japan in the 1960s, but it was not until the bubble economy years of the 1980s that funds began pouring in. At this time, corporations accounted for the lion's share of investing, spending ¥20.5bn in 1989, according to the Ministry of International Trade and Industry. Banks invested ¥15.8bn in ventures that year. As the bubble deflated, investor enthusiasm for venture businesses flagged: banks and corporations were strapped for cash. Funding dropped off sharply through the 1990s and did not pick up again until 1998. However, by 1998, the situation had changed dramatically. Instead of bank employees and corporate salarymen, this time around it was young mavericks from two new Japanese companies that were investing. Softbank and Hikari Tsushin altered the course of Japanese venture capital single-handedly by sending dozens of employees out on the street to look for deals. Softbank turned up 450 such companies; Hikari Tsushin does not disclose the size of its portfolio. See *Financial Times*, (http://specials.ft.com/ln/ftsurveys/industry/sc23436.htm) for the details.
5. According to the AVCJ (2000), the data contained in the survey are reliable but cannot be guaranteed to be correct and complete.
6. Detailed lists of industries which fall into each of these categories are presented in the Appendix.
7. See http://it-taskforce.nic.in/vsit-taskforce/bgr 3.htm.
8. VC employ a number of monitoring and information tools to scruitnize investee forms before providing them with capital. Afterwards the investee firms are monitored very closely. The monitoring and information tools of venture capitalists include meting out financing in discrete stages over time; syndicating investments with other venture capital firms; taking seats on a firm's board of directors; and compensation arrangements including stock options. For a detailed survey of a number of studies documenting the efforts of VCs in employing these tools, see Gompers and Lerner (2001).
9. The analysis here is restricted to 1999 as we do not have comparative data for the reference countries (Australia, Japan, Israel and the USA) for more recent years.
10. The minimum holding period (in years) of the equity investments by a venture capitalist in an investee company. Normally the lock-up period is three years from the start of investment. Only after the completion of this lock-up period is the venture capitalist allowed to divest off his/her holdings in a specific invested company.
11. However, data on the total number and amount of IPOs in five countries, Malaysia (since 1998, http://www.klse.com.my/website/listing/ipo1998.htm*)*, Hong Kong (since 1994, http://www.hkex.com.hk/listedco/newlist/1994.xls*)*, Singapore (only for 2002, http://info.sgx.com/webipo.nsf/IPO+By+Closing+Date?OpenView), Taiwan (only for 1999, http://www.tse.com.tw/plan/factbook/2000/table2.htm) and S.Korea (from 1999 onwards, http://www.kse.or.kr/eng/list/ncop/listNewCorp.jsp) are available. But in most cases the data merely refer to the amounts offered and not amounts actually subscribed.

12. Analysing the Indian situation with respect to the availability of VC professionals, McKinsey, the management consultancy firm, said: 'We have reviewed what's going wrong here [in India] and one issue recurs: we just can't hire quickly enough. To get world class valuations, we need world class people. But we can't find enough talented leaders to start and run a company. This is the biggest single barrier [for VCs].' See *Financial Times*, http://specials.ft.com/ln/ftsurveys/industry/sc23446.htm.
13. The choice of the year is dictated purely by the availability of data.

REFERENCES

Asian Venture Capital Journal (AVCJ) (various issues), *The Guide to Venture Capital in Asia*, Hong Kong: Asian Venture Capital Journal.

Aylward, Anthony (1998), 'Trends in Venture Capital Finance in Developing Countries', IFC discussion paper no. 36, http://www.ifc.org/economics/pubs/dp 36/dp 36.pdf.

Bank of England (2001), 'Financing of Technology-Based Firms', http://www.bankofengland.co.uk/hightech2001.pdf.

Bartzokas, Anthony (2001) 'Financial Markets and Technological Change: Patterns of Technological and Financial Decisions by Manufacturing Firms in Southern Europe', UNU/ INTECH discussion paper, 2001–4.

Bartzokas, Anthony and Morris Teubal (2002), 'Introduction to a special issue on the political economy of innovation policy implementation in industrialising countries', *Journal of the Economics of Innovation and New Technology*, **XI**(3&4).

Baskar, Vijay M. and Girija Krishnaswamy (2002), *Determinants of Dynamism in Software Clusters in Developing Countries: A Study of Bangalore, Southern India*, Bangalore: Indian Institute of Information Technology.

Black, B. and R. Gilson (1998), 'Venture capital and the structure of capital markets: banks versus stock markets', *Journal of Financial Economics*, **47**, 243–77.

British Venture Capital Association (2002), *A Guide to Venture Capital*, London: British Venture Capital Association.

European Commission (2000), 'Funding of new technology-based firms in commercial banks in Europe' (http://www.cordis.lu/finance/src/publicat.htm).

Gompers, Paul and Josh Lerner (2001), 'The venture capital revolution', *Journal of Economic Perspectives*, **15**, 145–68.

Hall, Bronwyn (2002), 'The financing of research and development', *Oxford Review of Economic Policy*, summer.

Hall, Bronwyn and John Van Reenen (2000), 'How effective are fiscal incentives for R&D? A review of the evidence', *Research Policy*, **29**, 449–69.

Hurwitz, Seth L. (1999), 'The Japanese Venture Capital Industry', MIT Japan Program, Working Paper Series, MITJP 99–04.

International Finance Corporation (1999), *Emerging Stock Markets Factbook 1999*, Washington, DC: International Finance Corporation.

Jeng, Leslie A. and Philippe C. Wells (2000), 'The determinants of venture capital funding: evidence across countries', *Journal of Corporate Finance*, **6**, 241–89.

Kortum, Samuel and Josh Lerner (2000), 'Assessing the contribution of venture capital to innovation', *RAND Journal of Economics*, **31**(4), 674–92.

Kuemmerle, Walter (2001), 'Comparing catalysts of change: evolution and institutional differences in the venture capital industries in the U.S., Japan and

Germany', in Robert A. Burgleman and Henry Chesbrough (eds), *Comparative Studies of Technological Evolution*, New York: Jai, pp.227–61.

Mani, Sunil (1997), 'Financial barriers to domestic technology development and venture capital: evidence from India', *Science, Technology and Development*, **15**(2&3), 227–40.

Mani, Sunil (2000), 'Exports of High Technology Products from Developing Countries: Is it Real or a Statistical Artifact?', Institute for New Technologies, United Nations University, Maastricht, discussion paper (http://www.intech.unu.edu/publications/index.htm).

Mani, Sunil (2002), *Government, Innovation and Technology Policy, An International Comparative Analysis*, Cheltenham UK and Northampton, MA, USA: Edward Elgar.

Mayer, Colin (2002), 'Financing the New Economy: financial institutions and corporate governance', *Information Economics and Policy*, **14**(2), 311–26.

Miller, Robert R. (2000), *Venture Capital in a Transition Economy – A Case Study of Three Hungarian Funds*, Washington, DC: International Finance Corporation.

National Science Board (2000), 'Science and Engineering Indicators: 2000', NSB 00-1, Arlington, VA.

OECD (1998), *Government Venture Capital for Technology-based Firms*, OCDE/GD(97)201, Paris: OECD.

Pfeil, Andreas (2001), 'Venture capital: new ways of financing technology innovation', background paper to the Human Development Report 2001, UNDP, New York.

Rausch, Lawrence (1998), 'Venture Capital Investment Trends in the United States and Europe', Division of Science Resources Studies, Directorate for Social, Behavioral and Economic Sciences, NSF 99-303.

Schertler, Andrea (2001), 'Venture Capital in Europe's Common Market: A Quantitative Description', working paper no. 01-4, EIFC Technology and Finance Series, UNU/INTECH, Maastricht.

Singh, Ajit, Alaka Singh and Bruce Weisse (2000), 'Information technology, venture capital and the stock market', background paper for the World Employment Report 2000–2001, International Labour Organisation, Geneva.

Storey, David J. and Bruce Tether (1994), 'A review of empirical knowledge and an assessment of statistical data on the economic importance of new technology based firms (NTBFs) in Europe' (http://www.cordis.lu/eims/src/eims-r 31.htm).

Storey, David J. and Bruce Tether (1998), 'Public policy measures to support new technology-based firms in the European Union', *Research Policy*, **26**, 1037–57.

Van Osnabrugge, M. (2000), 'A comparison of business angel and venture capitalist investment procedures: an agency theory-based analysis', *Venture Capital: An International Journal of Entrepreneurial Finance*, **2**(2), 91–109.

Venture Economics (1998), *Exiting Venture Capital Investments*, Wellesley: Venture Economics.

APPENDIX 1: INDUSTRY CLASSIFICATION ACCORDING TO THE *ASIAN VENTURE CAPITAL JOURNAL*

Agriculture/Fisheries

Agriculture/farming
Animal husbandry
Fishing
Forestry
Plantations

Computer-related

Computers – components
Computers – desktops, related equipment
Computers – mainframes
Computers – portable
Computers – retail
Peripherals
Semiconductors
Software – OS/other applications
Software – services

Conglomerates

Holding companies
Trading companies

Construction

Building materials
Building services and systems
Commercial development
Construction – other
Contractors – general and special
Residential development

Consumer Products and Services

Food and beverage
Food processing
Household products
Personal care products

Ecology

Alternative energy
Pollution control and recycling
Waste management

Electronics

Batteries
Consumer electronics
Electronics – other
Instruments
Lighting
Power supplies
Toys and games

Financial Services

Asset management
Commercial banking
Insurance
Investment banking
Leasing
Real estate
Securities – broker/dealer
Services – other financial
Venture capital/private equity

Information Technology

Internet – content provider
Internet – e-commerce
Internet – infrastructure
Internet – networking
Internet – offline services
Internet – online services

Infrastructure

Airports
Pipelines
Ports
Power generation
Roads and highways
Transportation – other

Leisure/Entertainment

Leisure/entertainment
Movie theatres

Manufacturing – heavy

Automobiles – cars
Automobiles – components
Building materials and products

Ceramics, clay and stone
Chemicals
Fabricated metal
Glass
Heating and cooling systems
Industrial machinery
Manufacturing – other heavy
Pulp, paper and packaging
Steel
Synthetics and rubber
Transportation equipment

Manufacturing – light

Furniture
Leather products
Lumber and wood products
Manufacturing – other light
Materials – other
Office equipment
Precision instruments
Tobacco

Media

Information services
Magazines and periodicals
Movies – production
Newspapers
Television and radio stations
Theatres

Medical/Biotechnology

Diagnostic/therapeutic products
Doctors and services
Drugs – OTC/prescription
Health science
Home healthcare
Hospital management
Pharmaceuticals

Mining and Metals

Iron and coal
Natural resources – other
Oil and gas – exploration and development
Oil and gas – refining and retail
Precious metals

Retail/Wholesale

Automotive
Books and printed material
Clothing and apparel
Eating and drinking
Electronics
Energy – gas, oil and alternative
Food and drugs
General merchandise
Speciality retail – other
Wholesale – consumer
Wholesale – trade

Services, Non-financial

Advertising/public relations
Business services
Consulting
Educational
Engineering
Legal
Medical
Personal services
Services – other non-financial
Temporary help

Telecommunications

Cable
Cellular and wireless
Phones and related equipment
Satellite
Service provider

Textiles and Clothing

Textiles – manufacturing
Textiles – raw materials

Transportation/Distribution

Air cargo
Airlines
Buses
Cabs and cars
Couriers
Pipelines
Railroad
Rockets and orbital

Shipping
Trucking
Warehousing

Travel/Hospitality

Airport services
Hotels and lodging
Restaurants and pubs
Tourism services – other
Travel agencies

Utilities

Electric
Gas
Water

7. Financial systems, investment in innovation, and venture capital: the case of China

Steven White, Jian Gao and Wei Zhang

INTRODUCTION

The Chinese government has always seen science and technology as essential in supporting its ambitions for national security and, more recently, economic development. Venture capital in the Chinese context, therefore, has been promoted not as a means to private gain, but as a critical mechanism for linking scientific and technological capabilities and outputs, on the one hand, with national and regional economic and social development, on the other.

No longer, however, do policy makers or analysts ask the naïve question of whether China's venture capital industry will follow the 'Silicon Valley model', that of some other country or region, or develop into a distinctive 'Chinese' model. Although still developing, China's venture capital industry is clearly an outcome of its particular combination of political, economic and social institutions[1] and the nature of the broader changes it has been undergoing during its transition from central planning to a more market-based business system.

China's venture capital industry, including the total set of related actors and institutions, has undergone a dramatic transformation over the last two decades. Because of its starting conditions – in particular, its legacy of inefficient central planning and socialist ideology – the results of this transformation seem particularly striking. The system that has emerged so far is highly complex in terms of the variety and number of organizational actors, as well the multiple dimensions on which these actors are linked (Figure 7.1). This complexity is increased because all of the organizational and institutional elements are themselves changing in response to policy, technological and other developments.

Although still in flux, the system has already generated impressive results in terms of sheer scale (see Table 7.1). It now includes 86000 new technology-based ventures employing 5.6 million and generating revenues

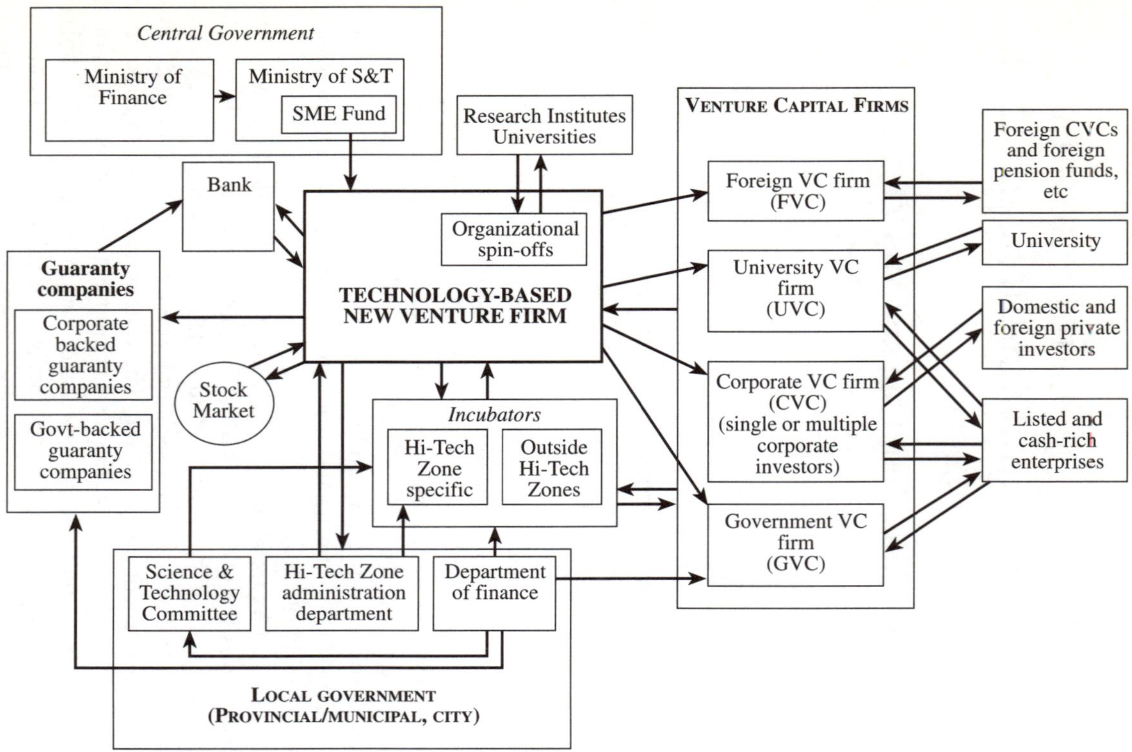

Figure 7.1 Actors and flows in China's VC industry

Table 7.1 Status and performance of elements of China's VC industry[1]

Technology-based new ventures[2]
approximately 86000 nationwide (8.3% annual increase)
employ 5.6 million (14% annual increase)
combined revenues of RMB1.5 trillion (40% annual increase)
reported profits of RMB100 billion, remitted taxes of RMB78 billion (39% annual increase)
exported RMB214 billion (US$26 billion; 64% annual increase)
spent RMB41 billion on R&D (5.7% annual decrease)
main categories of ownership form: 7.3% state-owned enterprises (SOEs), 16% collectives, 24% private/individual firms, 6% joint stock companies, 37% limited liability companies, 5.5% foreign-invested (from other countries, Hong Kong, Macau and Taiwan)
Venture capital firms
160 domestic VC firms, 50 foreign VC firms
no new VCFs founded since July 2001
several have gone bankrupt
Science and technology industrial parks, high-tech zones
53 nationwide
housing 20800 high technology-based firms (primarily technology-based SMEs)
employ 2.51 million
combined revenues of RMB921 billion
taxes and profits of RMB106 billion
exports of RMB153 billion (US$18.6 billion)
location of approximately half of the projects funded by the Torch Program
Incubators
465 technology business incubators nationwide
house 788 intermediary service agencies
15449 tenant firms, employing 292000, including:
130000 with bachelor, master or PhD degrees
4100 returnees from study abroad
combined sales income of RMB42 billion
3887 firms had 'graduated' and moved out of the incubators
32 of these listed on the stock market
Torch Program
Torch Program projects
approximately 1000 projects selected each year
funded by bank loans or self-raised funds
2870 national-level projects had sales of RMB83 billion, taxes and profits of RMB15 billion and exports of RMB16.5 billion (US$2 billion).
Innovation Fund for Small Technology-based Firms
2900 projects funded during 1999–2002
RMB2.3 billion disbursed (supplemented by RMB22.5 billion from local governments, banks, VC and so on)

Notes:

1. As of the end of 2000, unless otherwise noted.
2. Small and medium enterprises in which at least 30% of the employees are S&T personnel, and R&D expenditures represent at least 5% of sales.

of RMB1.5 trillion (US$182 billion). Supporting them are over 200 venture capital firms, at least 130 publicly listed firms, 465 technology business incubators, and 53 high-tech zones, as well as the central bureaucracies and provincial and municipal governments.

This chapter examines the development, structure and key issues facing China's venture capital industry and its impact on new technology ventures. Although it is by no means stable in terms of structure or dynamics, we are able to discern certain trends and trajectories. We are also able to understand the nature of the outcomes of this system to date regarding the development of new technology ventures, various forms of venture capital firms and organizational structures supporting them. We relate these features of the system to indicators of its performance, as well as the current and emerging issues affecting its further development.

HISTORICAL EVOLUTION OF CHINA'S VC INDUSTRY

Venture capital represents a set of major institutional and organizational changes in China, especially compared to the central planning system operating into the 1980s. The timeline in Table 7.2 presents the major policy and business developments related to China's VC industry since 1984. Although the result has been dramatic, the series of changes are best seen as evolutionary and primarily driven by China's larger objective of national technological and economic development. Key changes in China's science and technology policy and business system structure during the transition era can thus be linked to the emergence and nature of China's venture capital industry. In this context, venture capital is simultaneously an extension of prior policy trajectories and a potential answer to problems that other policy initiatives have not been able to solve.

Three changes during the transition from central planning to the (still evolving) market-oriented business system[2] represent the principal policy and structural antecedents for China's venture capital industry. First, a growing number of policy makers and ministries have developed an increasingly sophisticated perception of cause-and-effect relationships linking technological and economic development. By the late 1970s, pragmatic leaders had recognized the inefficiencies and lower effectiveness of a centrally planned economy in practice. The R&D system in China under central planning and lasting until the early 1980s comprised a large number of organizations specializing in particular industries and in specific stages of the value chain (research, development, manufacturing, distribution and so on), each answerable to one or several parts of the central and often local

Table 7.2 Key developments in China's VC industry

	Government policy and regulations	Enterprise and business
1984	National Research Center of Science and Technology for Development (under State Science and Technology Commission, SSTC) first organizes research effort on 'New Technology and China's Countermeasures', and suggests that a venture capital system be established to promote the development of new and high technology.	
1985	The Chinese Communist Party (CCP) and State Council release 'The Decision on the Reform of the Science and Technology System' that notes that venture capital could be used to support high-tech R&D in areas of rapid change and high risk China's first patent law enacted	SSTC and Ministry of Finance (MoF) establish and fund China New Technology Venture Investment Corp., the first limited corporation in China focused on venture capital
1986	863 High-Tech Program started (10 years, over RMB10 billion in funds for scientific research)	
1987		China's first incubator established by local government in Hubei: Wuhan East Lake Entrepreneur Service Center
1988	Torch Program launched to promote spin-off ventures from research institutes and universities, including the government's direct investment	
1989	State Council and Ministry of Foreign Trade and Economic Cooperation (MOFTEC) permit China Merchants Holding (Hong Kong), SSTC and Commission of S&T and Industry for National Defense to establish Kezhao High-Tech Ltd, China's first Sino-foreign joint venture in venture capital, with the mission of funding the industrialization of R&D results from national high-tech plans (863, Torch)	
1990		

Table 7.2 (continued)

	Government policy and regulations	Enterprise and business
1991	State Council announces 'Authorization of National High-Tech Zones and Related Policies', allowing relevant departments to set up VC funds in high-tech zones to support risky high-tech industry development, and mature high-tech zones could set up VC corporations	SSTC, MoF and Industrial and Commercial Bank of China establish the Technology Venture Development Center
1992		Technology Venture Development Corporations established by local governments in Shenyang, Shanxi, Guangdong, Shanghai, Zhejiang
1993	Standing Committee of the National People's Congress (NPC) approves 'Science and Technology Promotion Law of China'	ChinaVest invests in Zindart, a new venture (but not high-tech) that listed its American Depository Receipts (ADRs) on NASDAQ in 1997
1994		The Pacific Technology Venture Investment Fund of the US firm International Data Group (IDG) establishes three VC companies with the local S&T commissions of Beijing, Shanghai and Guangdong
1995	CCP and State Council announce 'The Decision on Accelerating Scientific and Technological Progress', emphasizing the development of VC and establishing a technology venture capital system	
1996	State Council policy document 'On Further Improving China's S&T System' emphasizes the need actively to investigate and develop a VC system to promote China's S&T outputs National People's Congress passes 'Law Promoting the Industrialization of China's Technological Achievements', the first legal statement allowing VC as a commercial activity and funds to be raised from national	At least 20 VC firms established by S&T commissions and finance departments of local governments

Table 7.2

	Government policy and regulations	Enterprise and business
	or local governments, enterprises or other organizations, or individuals to support technology ventures SSTC sends delegation to USA to study laws and policies related to small company investment, intellectual property rights, and VC, and results discussed at meeting of local science commissions along with heads of finance departments of People's Bank of China, four other state-owned banks, and other bureaus of the SSTC	
1997	Deng Nan (daughter of Deng Xiaoping and Vice Minister of SSTC) appointed to oversee study of VC system, and directed a group from the School of Economics and Management of Tsinghua University to deliver a report with practical recommendations for a VC system structure, the relationship between VC and capital markets, and plan for establishing a VC system 973 Program (RMB4.5 billion) initiated to support basic research	China's first VC firm, founded in 1985, was declared bankrupt and closed by the People's Bank of China Zindart, a toy manufacturer that received investment from ChinaVest in 1993, is first new Chinese venture to list ADRs on NASDAQ New tech-based venture AsiaInfo receives US$18 million investment from three foreign VC firms Sohu.com predecessor firm receives US$6.5 million investment from foreign VCs, and is first new venture in China's IT industry to receive investment
1998	Prime Minister Li Peng chairs a meeting of China's leading policy group on S&T, concluding that a general plan for a VC system be made and implemented Vice Prime Minister Zhu Rongji forms a coordination group including the State Planning Commission, People's Bank of China, China Securities Regulatory Commission and relevant government departments, supported by the finance	Sohu receives US$2.2 million investment from foreign VCs. Kingdee receives RMB20 million investment from Guangdong Pacific Investment Corp. and joint investment by IDG and Guangdong's S&T Bureau. IDG signs cooperation

Table 7.2 (continued)

	Government policy and regulations	Enterprise and business
	research centers of the Academy of Social Sciences and the Bank of China Deng Nan (Vice Minister of the Ministry of S&T (MoST), formerly SSTC) discusses VC system and mainland high-tech firm listings with president of the Hong Kong Stock Exchange After vetting with the Education Commission and Finance Commission of the NPC, MoST submits 'Report on Establishing China's S&T Venture Capital System' to State Council 'No.1 Proposal' on developing China's VC industry by the Central Committee of the Chinese National Democratic Constructive Association, presented at the Ninth Conference of the NPC, creating a wave of VC firm foundings, including local governments' direct investments in VC firms	agreement with MoST for IDGVC to invest $1 billion over seven years in Chinese new high-tech ventures and promote Chinese high-tech industry About 90 VC firms active in China, with RMB7.4 billion under management
1999	Prime Minister Zhu Rongji accepted final report of MoST, but directed that S&T VC should primarily support small and medium-scale enterprises (SMEs) Group formed by NPC to draft a VC law, on which seven ministries would provide input and opinion before the 'Procedure for Managing the Industrial Investment Fund' was debated by the CCP and State Council and supported in their document 'Decision on Strengthening Technological Innovation, Developing High-Tech and Realizing its Industrialization' First international discussion held regarding the drafting of the Investment Fund Law Technology-based SME Innovation Fund established, overseen by MoST	
2000	Shenzhen enacts the first local regulatory statutes for VC: 'Temporary Regulations for VC Investing in High-Tech Industry in Shenzhen' NPC holds second international meeting to discuss the Investment Fund Law	Beijing VC Association (formed in 1999) formally registered with government, followed by associations in Shenzhen and Shanghai First incubator funded by a

Table 7.2

	Government policy and regulations	Enterprise and business
	State Council announces 'Policy for Encouraging the Software Industry and Promoting the IC Industry'	private enterprise, Jinghai Business Incubator, established in Zhong guancun Science Park AsiaInfo and UTStarcom become first Chinese tech-based new ventures to list shares on NASDAQ, followed by Sohu, Sina.com and Netease
2001	Technology-based SME Innovation Fund (est.1999) disburses RMB1.96 billion to 2577 projects by end of 2001 Beijing enacts its VC regulations ('Byelaw of Zhongguancun Science Park'), regulating VC operation, organizational structure, registered funds and means of return. Also releases 'Management of Limited Liability Corporations (no.69)' to promote and regulate the development of limited VC corporations in the Zhongguancun Science Park MOFTEC, MoST and the National Industry and Commerce Administration release and enact the 'Temporary Regulations for Establishing Foreign Venture Capital Corporations' VC Investment Committee of the S&T Finance Promotion Association, a semi-government organization, is established in Beijing and is the first truly cross-regional, national organization focused specifically on VC, with mission to promote linkages between government and private VC, study government environment for successful VC industry, exchanges within the VC industry, consolidate activities and experience, and develop training. Currently 132 companies and 160 individual members	First limited partnership VC corporation in China established in Beijing (Beijing Tianlu VC Center), a joint venture of Tianye Corporation and the Economic Construction and Development Corporation, both of Xinjiang, and Beijing's Sinotrust, with the chairman and President of Sinotrust appointed as CEO; closed the same year. 465 incubators registered nationwide, funded by government, universities, research institutes, SOEs, private and foreign enterprises Kingdee becomes first Chinese high-tech venture to be listed on Hong Kong's new Growth Enterprise Market (GEM)
2002		Approximately 160 domestic and 50 foreign-funded VC firms active in

Table 7.2 (continued)

Government policy and regulations	Enterprise and business
	China, but no new VC firms founded since July 2001 and investment activity considerably slowed China Venture Capital Association (CVCA) registered in Hong Kong, and now includes over 50 VC firms with a total of US$60 billion in funds and annual investments of $300–500 million in Greater China

government, but with few or no horizontal linkages and information flows between these functionally specialized actors (Liu and White, 2001). Innovation was primarily initiated by central government ministries and bureaus, within the guidelines of the State Planning Commission's (SPC) national plans. These bureaucracies claimed both authority and responsibility for initiatives such as technological development, adoption, upgrade or transfer, as well as for definition of production output and distribution.

Accordingly, science and technology development efforts were driven by policy objectives, and from 1956 the priority was on developing research and production capabilities in atomic energy, electronics, semiconductors, automation, computers and rocket technology (McDonald, 1990), and massive resources were also injected into ultimately successful mission projects, developing atomic and hydrogen bombs (1964 and 1967, respectively) and launching satellites (1967). This system structure and innovation dynamics stand in stark contrast to those of most developed countries, in which large firms play a central role in R&D, undertake activities across the value chain (although certainly not completely independently) and make their own decisions regarding science and technology initiatives, funding and outputs.

The complete dependence on the central government for identifying and providing resources for innovation, however, created an environment in which there were no incentives for organizational actors in the innovation system (research institutes and manufacturers, in particular) to introduce, adopt or diffuse innovations proactively. There was no market competition, profit or other operational efficiency-based criterion for their performance.

In this context, many of the reforms introduced in the 1980s can be interpreted as the central government's willingness to experiment with a new institutional structure that would be more effective at introducing, transferring and exploiting new technology. The emergence of a few successful entrepreneurs (for example, Stone Electronics) also spurred the central and local government to re-examine its science and technology (S&T) policies and related institutions.

The first wave of reforms in the 1980s had two objectives. The first was to increase science and technology outputs, reflecting the assumption that a greater supply of technology would lead to greater diffusion and implementation, which would support both technological and economic development objectives. During the same period, and as part of the larger reforms under way in China from the early 1980s, the government began to shift responsibility and, more gradually, authority for resource allocation decisions from the central government bureaucracies to the operational organizations (research institutes, manufacturers and others). This was accompanied by increased responsibility and alternatives for improving financial performance, either by generating or increasing revenues or by winning competitive project funds from the government (Naughton, 1994; Child, 1994).

These reforms, however, were not as successful as hoped in terms of bringing new products to market and improving the technological base of China's (predominantly) state-owned industries. Policy makers began to realize that generating more S&T outputs was not enough. Increasingly, policymakers became aware of the importance of linkages across functional activities and stages of research, development and manufacturing (White and Liu, 1998, 2001; Liu and White, 2001). Prior reforms that decentralized most resource and operational decisions had at the same time largely removed the central government as the mechanism linking functionally specialized organizations. Studies of the innovation process in other contexts, however, emphasize the critical importance of linking activities and resources across the value chain. Successfully managing such linkages is a challenge, even within the same organization. In the Chinese context, where linkage had to occur across organizational boundaries, it was even more difficult. This realization led the government to pursue both cross-organizational and internal solutions to this problem of linkage. Perhaps the first major initiative was in 1985 when the government encouraged the establishment of technology markets (auctions in which research results are traded). This was followed by encouraging state-owned manufacturers to establish R&D centers internally, promoting mergers between S&T organizations and manufacturers, and allowing S&T institutes to undertake manufacturing operations to exploit the technology they developed.

By the end of the 1980s, however, it became clear that only the technology markets and functional diversification of S&T institutes were having a significant impact. Mergers of research institutes and manufacturers have been extremely rare, and those that have been implemented have been conflictual and the results disappointing. One problem was that cash-strapped SOEs were not able to support the additional financial burden of a research institute (Gu, 1999). Establishing R&D centers and capabilities within manufacturers has been primarily window-dressing, with little evidence that such centers have actually contributed significantly to innovative capabilities and outputs, beyond simple quality control or analysis. The general lack of related managerial experience and expertise has been a major barrier to such centers invigorating innovation within SOEs (Gao and Fu, 1996).

Although technology markets, first initiated in 1985, have grown dramatically by all measures (numbers of markets operating, transactions, value) they have primarily encouraged technical consulting relationships, rather than the transfer of S&T results (technology transfer) or joint development. The instances of these technology market transactions involving truly collaborative activity across organizational boundaries have been rare, however, and even more so when the technology is closer to commercialization. Gu (1999) emphasizes the difficulty of market-type exchanges in transfers of resources in which there is a high degree of uncertainty, and this is particularly true of joint development relationships in which often the outcome, and even the process, cannot be specified ex ante. Indeed this is one factor explaining the different strategies that SOEs have used to acquire resources with different degrees of uncertainty (White and Liu, 2001). The lack of absorbtive capacity (Cohen and Levinthal, 1990) on the part of most SOE recipients also hinders the use of technology markets in transferring technology or collaborative R&D. Indeed, the degree of internal R&D capabilities was a key factor predicting joint development as against simple technology transfer in China's pharmaceutical industry (White, 2000).

In addition to the sources of difficulty in managing such relationships in any context, collaboration across organizational boundaries has been severely hampered by China's inadequate institutional environment for governing such relationships. This includes an immature corporate legal code – especially contract law – and its uneven enforcement, and ambiguity regarding intellectual property rights; that is, which individuals or organizations have legitimate claim over what type of property rights. After the government bureaucracies withdrew from their central planning era roles as governors of interorganizational relationships, S&T organizations and enterprises had to rely on formal contracts and trust (whatever the source of that trust; see, for example, Zucker, 1986). As long as actors perceive that there is a high risk of opportunistic behavior by a partner, with no effective legal or other safeguards, they will rationally try to avoid such relationships. This helps explain why the instances of interorganizational collaboration have increased in the area of academic research (as seen in the increase in number of papers being published by collaborators in different types of organization), but a simultaneous decrease in joint patenting over the same period (Liu and White, 2001).

These disincentives and barriers to collaboration and technology transfer, and the economic potential for significant revenues from exploiting the technology internally, could be argued to have created too strong incentives for the institutes to implement new technology themselves. These research institutes and universities lacked financial resources and capabilities in manufacturing, sales and distribution, and these resource and capability gaps could result in the technology being underexploited (ibid.).

By the mid-1990s, central government policy makers still had not found an effective solution to the basic objective that had initiated China's reform period in the late 1970s. Although the terminology they used to discuss these issues had been loosened from communist dogma, leaders were still searching for the best means to develop and derive economic value from new technology to support national developmental goals. Furthermore their basic policies, institutions and practices to promote new technology ventures had not changed significantly since the mid-late 1980s.

Thus the system that emerged in the 1990s consisted of three primary institutional actors providing resources for these ventures. First, R&D institutes and universities played the primary role at the start-up stage, providing both the original technology and seed capital for the venture. The Torch Program's actual financial contribution to these new ventures was relatively minor. Instead the primary benefit of being designated as a Torch Program project was the signal it gave to banks to provide loans to these ventures for technology commercialization.

The technology that was the substance of these projects was typically embodied in the spin-off of an entire institute, one of its subunits or a group of individuals. Various estimates suggest that such institute-initiated new ventures represented approximately half of such ventures operating in technology zones, or over a thousand such ventures, by 1993 (Gu, 1999:83). The source institution, using its new authority to allocate resources, would also provide financial support. For example, of the new technology enterprises founded in Beijing, an average of 85 per cent of their start-up capital came from the originating institution.

Accordingly the next set of actors who played an important role in new technology ventures were the banks, who were the primary source of financing. They, rather than the government, provided the majority of the investment in spin-off projects under the Torch Program. Although representing only 10 per cent of that investment in 1988 when the Torch Program began, their share increased to 50 per cent by 1990 and 70 per cent by 1991 (ibid.: 352). The banks themselves did not have the capability or access to critical information to assess risk at this initial start-up stage. Instead, they relied on a project's designation as a recipient of Torch Program support as policy guidance. The majority of bank financing, however, was available only at the expansion and later stages of the new ventures, with local governments acting as guarantors. Even into the mid-1990s, banks were the main financers of new venture expansion, but essentially absent as financers at the seed capital and start-up stages of these ventures.

Technology zones were the third source of support for new ventures, officially sanctioned in 1991 and extending a local experiment by the Wuhan government in 1987. These became a key source of support for new technology ventures. Gu (1999) has described them as an institutional interface between the new ventures and the broader, and in some ways inadequate, socioeconomic system into which the ventures were founded. First, they provided incubator functions, including physical space and infrastructure. Second, they licensed the new ventures in order for them to qualify for preferential treatment under the Torch Program and other government policies, and to gain access to funding from various sources, especially banks and venture capital firms. Local governments supported them because, by locating in these zones, the new ventures were seen as contributing to local economic development.

Venture Capital as a Solution to New Venture Investment Bottleneck

The policies, institutions and actions over the 1980s and early 1990s resulted in a large number of new technology ventures being founded in China before a separate venture capital industry and related institutional regulations were established. However, by the mid-1990s, central government leaders recognized that the current system for establishing new ventures, as a means of pursuing broader national developmental objectives, had reached its limits. There were several features of the current system that led to that view. First, the supply of initial-stage seed capital was too small in effect, dependent as it was on the very limited resources of research institutes and universities. Banks were strapped by their non-performing loans, and increasing loans to inherently high-risk ventures was untenable. Similarly, neither the central nor local governments had the surplus funds to step in as alternative financers of new ventures. An institutional bias against financing individual private ventures also represented a barrier to possibly promising new ventures being established.

In addition to the limits of the existing system to finance new ventures, the system also failed to provide the legal and institutional support necessary to channel available funds to new ventures. The government did not recognize venture finance organizations (that is, venture capital firms) as a legitimate organizational type. Until it did, such financing was either internal, as the institutes and universities allocated their own resources to new ventures, or a category of central or local government funding, whether through the Torch Program, zone incubators, or other funds aimed at new technology venture support. Although the China New Technology Venture Investment Corporation was formed in 1986, it was founded by the State Science and Technology Commission and the Ministry of Finance and operated as an SOE. As such, it was essentially a central government agency with the mandate to support national technology venture policy objectives, rather than a profit-oriented private enterprise.

A broader issue that was particularly relevant for new venture investment was the lack of an adequate legal framework and enforcement to enable new types of investors to provide financing to new ventures. This same problem has already been cited as one reason for the rarity of truly collaborative development activity between organizations, and the failure of technology markets to encourage research institutes to transfer commercially promising technology to enterprises. If parties do not have confidence in institutional safeguards, such as contract law, they will rationally avoid exposing themselves to the resulting risk of a transaction. This situation was exacerbated by the generally murky state of property rights in China regarding who has what rights over the use, rent extraction and transfer of assets (Steinfeld, 1998). Venture capital, defined as high-risk equity investment, is not possible if there is no legal definition and protection of ownership over a new venture's assets.

Gradually, from the mid-1990s, the perception of venture capital shifted from its being a type of government funding to being a commercial activity necessary to support the commercialization of new technology. Foreign VC firms had already been allowed to register as commercial enterprises in China in the 1980s, although their investment activities were extremely limited by the lack of suitable investment projects.[3] The founding of domestic VC firms began with the establishment of local government-financed venture capital firms (GVCFs), followed by university-backed VC firms (UVCFs). With Announcement No.1 at the Ninth Conference of the NPC in 1998, however, corporate-backed VC firms could be established, and there was a wave of foundings involving government, corporate and foreign capital.

From that point, venture capital shifted from being a topic of policy research, discussion and experimentation, or a form of government subsidization of new technology ventures, to being a rapidly growing segment of

China's commercial financial system. The next section describes the structure of today's VC system that has emerged in the years since venture capital was formally sanctioned. It identifies the key actors and their relationships in the context of founding and financing new technology-based ventures in China.

STRUCTURE OF CHINA'S VC INDUSTRY

The venture capital industry that has emerged from China's policy, structural and institutional trajectories described in the preceding section is represented in Figure 7.1. As a system, it represents the current, albeit evolving, 'solution' in China for funding and nurturing technology-based new venture firms. It is diverse and complex in terms of both types of actors and the diversity of interactions among them.

Because many of these differ significantly from the structure of venture capital systems in other national contexts, it is useful to describe the categories of actors and the nature of their relationships. The actors can be variously categorized, for example, distinguishing among them on the basis of ownership or control (for example, government bureaucracies, government-controlled organizations, and relatively autonomous organizations), or primary role vis-à-vis new technology ventures (for example, provider of financing, knowledge resources or political and social support). Because of the government's central role, we begin with a description of bureaucratic and government-linked actors in the system. We then describe the newer actors that have emerged, including different types of venture capital firms, as well as the new ventures themselves.

Government as Enabler

As the first section described and as is clear from the timeline of events (Table 7.2), the government has played a central role in the development of China's venture capital industry. The Ministry of Science and Technology (MoST, formerly the State Science and Technology Commission) was the primary champion, interpreting venture capital as a key factor behind the success of the high-tech industries in the United States. In the Chinese context, venture capital came to be seen as a means of linking science and technological development, on the one hand, with national economic development, on the other. During the transition period, MoST was able to garner support from other key central government bodies, including the State Council, State Planning Commission and the Chinese Communist Party leadership. This top-level support then led the way for bringing on board other important bureaucratic actors, in particular the Ministry of Finance and local governments.

The result of these developments has been a de facto division of labor between central government bureaucracies and local governments vis-à-vis new venture firms. The central government has played three important roles. First, its transition-era policy of decentralization of responsibility and authority has created the institutional space for lower level actors – both local governments and science and technology organizations (that is, research institutes and universities) – to act entrepreneurially and undertake new activities. The result for local governments is discussed below. This has allowed research institutes and universities to spin off organizational sub units, people and even whole organizations to become the basis of new venture firms.

The second important role has been to provide legitimacy to technological entrepreneurship as a commercial activity and to new ventures as legal entities. Thus the funds that have flowed directly from central government sources to new ventures, for example, serve a more important role as a signal to other actors rather than as a source of financing. Indeed these funds are more accurately described as 'leading funds', serving as a signal to local governments and banks that the venture is politically and socially legitimate and a qualified recipient of financial and other support.

Finally, the third important role of the central government has been to create an institutional environment conducive to technology-based new venture development. Although this is still under development, the government has made impressive strides towards aligning the legal and financial systems more closely to the goal of establishing a market-oriented business system. For venture capital and new venture firm development, key institutional elements include corporate law governing the status and activities of legal entities, investment, contracts and intellectual property; regulation of foreign capital and enterprises; and the stock market and other elements of the capital markets.

Compared to the central government, local governments have played a much more direct role in the development of new ventures and supporting infrastructure. Of course, one reason for local governments responding so positively to central government's initiatives in this area is the still considerable control of the central government over key rewards to individuals and organizations that stand out as supporters and implementers of policy initiatives. This is exercised through the still major role that the central government and Communist Party exert over upper level personnel appointments, both in government and in enterprises.

Broader support for central government initiatives within local governments, however, is motivated by the same fundamental objective as that of the central government: the consensus that greater exploitation of local science and technology resources can support economic and social development objectives, albeit focused on the local rather than national level. Indeed the local governments have an increasing incentive to pay attention to local economic growth, as central government support for both their budgets and local enterprises (especially SOEs) has dropped sharply, concomitant with an increase in authority for pursuing economic growth.

For these reasons – pursuit of recognition, searching for revenues and employment opportunities – local governments have responded enthusiastically to the incentives and opportunities to foster new technology-based ventures in their regions. More than the central government, the local governments, specifically the local departments of finance, bureaus of the science and technology committee, and high-tech zone administration departments, have provided a range of direct support to new ventures. For example, the departments of finance have created government-backed guaranty companies to guarantee bank loans to local ventures, in addition to direct financial support to new ventures. The local governments have also allowed these firms greater operating autonomy, including offering competitive compensation to their employees.

Local governments have also used high-tech zones and specific incubator organizations within these zones to support the development of new technology ventures. They provide various levels of support to new firms, including tax exemptions and reductions, physical space at low rental rates, leasing, better social services and other preferential conditions. In 1991, after the State Council authorized local governments to provide VC funds through these zone administrations, their role and activities increased dramatically. As a result, by 1992, there were already 52 high-tech zones established throughout China, 5569 new technology ventures were registered, and their combined output was estimated at RMB23 billion (Gu, 1999:39). By 2000, the number of new ventures in these zones had exceeded 20000 (Table 7.1).

The central government considers the zones to be a successful policy initiative. From MoST's perspective, these zones have generated benefits in two areas (Chen, 2002). First, they have provided the structure in which local governments can express their creativity by adopting and improving the policies and activities related to these zones. They have allowed experimentation in terms of administrative structure, market-oriented operations and human resource management, all in line with the overall thrust of economic reforms under way in China. Second, the zones have contributed significantly to the commercialization of China's S&T outputs by non-government S&T firms, as well as by serving as an important base for training and education. They have also helped enhance the competitiveness of these firms by supporting their continual innovation capabilities.

Coinciding with the development of S&T zones in the early 1990s, incubators emerged in force. They were first founded within zones as extensions of the original services provided by the zone administration within the local government. The 465 incubators registered nationwide are now found both within and outside zones, and receive funding from all of the sources backing VC firms. Indeed some of these incubators are even treated as a category of new technology venture firm by their investors. The outputs of these incubators are impressive; by 2000, nearly 4000 firms had emerged from them, including 32 that had been listed on the stock market (see Table 7.1).

The local government usually provides incubators, whether within or outside S&T zones, with physical infrastructure and favorable policies, such as those related to leasing space, tax incentives and basic services. A number go even further, acting as intermediaries and providing training and management services. This could even be to the extent of providing platform software services, although usually through a larger industrial firm with those resources and capabilities.

Now, anyone may establish an incubator as a for-profit firm. Beijing, with the largest number of incubators, has special policies for promoting incubators, regardless of their location vis-à-vis zones, and whether backed by the government, corporations or other private financing. There is a licensing process by which an incubator is authorized as such by the Beijing Science and Technology Commission, including standards of operation and assessment by a group of experts. Authorized incubators are re-examined every two years to confirm that they still meet these requirements.

Many zone or government-backed incubators are actually state entities, with many of the managers coming from the government. As a result, in many of these organizations, incentives are inadequate, nor do these managers have the expertise to provide strong support and expanded value-added services. University-based incubators are better than pure government-backed incubators in terms of both their internal systems and human resources. Corporate-backed incubators are even more strongly focused on creating profit and value than the other types. Of course, they are liable to an overemphasis on short-term profits at the expense of longer-term investment and development. Although university, corporate and purely private incubators may not be under direct government control, most still seek local government support, especially that related to physical space, infrastructure supply and tax incentives.

Venture Capital Industry Profile

The cumulative result of central and local government promotion of and direct involvement in venture capital has resulted in a dramatic wave of new venture capital firms, capital and investments (see Tables 7.3 and 7.4). By 2001, 246 venture capital firms were registered, reporting almost $5 billion in funds under management. They have become the primary source of funds for new technology-based ventures in China. Since the mid-1990s, domestic venture capital firms have accounted for the majority of disbursements, leveling off at approximately 78 per cent since 1999.

Table 7.3 Venture capital industry profile

	1997	1998	1999	2000	2001
VC firms	38	59	99	201	246
VC professionals (est.)	NA	NA	NA	NA	2214
Capital under management ($m.)	922	1262	2500	4527	4915
New funds raised ($m.)	485	340	1238	2027	388
New investments during year*		38	67	261	178

Note: * Domestic VC firms only.

The primary targets of these funds (87 per cent as of 2001) have been in areas categorized as high-tech (Table 7.5). Software attracted the largest percentage of funds (15.7 per cent). Several areas related to information technology (networking, telecommunication and other IT areas), however, together accounted for a larger percentage of funding (21.3 per cent). Other areas are also well represented, including pharmaceuticals and healthcare (9 per cent), biotechnology (8.5 per cent) and new materials (8.3 per cent). As will be discussed in the next section, domestic VC firms have favored investments in these areas, and they represent the majority of disbursements. Foreign firms, less biased towards high technology, account for a relatively larger proportion of investments in areas not considered 'high technology'.

China's venture capital industry has also seen a decline, although perhaps not as dramatic as in other countries, in overall indicators since 2001. Much less new capital was raised in 2001 ($388m. as against $2027m. in 2000), and investments declined by 32 per cent per year. By the end of 2002, the number of active firms was also estimated to have decreased to 200 or fewer.

Over the same time period, the industry has shown a dramatic change in the stage of disbursements (Table 7.6). As late as 1998, early-stage investments accounted for 81 per cent of disbursements, as against 19 per cent for expansion. By 2000, however, early stage investments had dropped by half as a percentage of total disbursements, and expansion and mature-stage investments have become the majority.

Table 7.4 Venture capital trends

	1991	1992	1993	1994	1995	1996	1997	1998	1999	2000	2001
Total VC capital pool ($m.)	67	113	231	267	364	437	922	1262	2500	4527	4915
New VC capital ($m.)		46	127	36	97	73	485	340	1238	2027	388
Actual new investment ($m.)								38	67	261	178
VC investments								70	116	395	257
Disbursements											
Domestic VCFs (%)				64	65	66	70	76	78	77	78
Foreign VCFs (%)				36	35	34	30	24	22	23	22

Table 7.5 Disbursements by industry (per cent)

Manufacturing	6.4
Agriculture	3.1
Traditional non-tech firms	3.1
High-tech	87.0
Software	15.7
Hardware	3.6
Networking	9.3
Telecommunications	7.5
Other sectors in IT	4.5
Semiconductor	0.9
Pharmaceutical and healthcare	9.0
Environmental protection	2.7
Biotechnology	8.5
New materials	8.3
Resource development	1.0
Opto-electronics	6.4
S&T services	4.0
New energy	3.4
Other key technologies	2.7

Table 7.6 Disbursements by stage (per cent)

	1998	1999	2000	2001
Early stage	81	54	41	42
Seed	14	13	13	12
Start-up	67	41	28	30
Expansion	19	41	54	50
Maturity	0	5	5	8

VC FIRM INVOLVEMENT IN NEW VENTURES

The nature of the relationship between venture capital firms and new ventures in China has varied considerably by the type of venture capital firm. Four distinct categories of VC firms, each with different antecedents, objectives and operating characteristics, have appeared in China. Table 7.7 presents a comparison of these four types, with examples of each presented in Tables 7.8a–7.8c. Each type of venture capital firm is evolving as their managers and backers learn through trial and error, as investees gain experience in working with them and as regulations and policies evolve. However, there are important trends within each of these types.

The first to appear in China were the government VC firms (GVCFs). Although the first such venture capital firm was established by the central government (SSTC and MoF) in 1985 (beginning operations in 1986), those that followed were all controlled by local governments, usually led by the local bureau of the science and technology commission and supported by the finance department of the local government. Although local governments were their initial source of financing, over time, and with changes in the regulatory environment, they have diversified their funding sources. Indeed they are increasingly dependent on listed and cash-rich enterprises to keep up their investment capacity.

GVCF's linkages to the local government and thereby new venture developments in those zones represent preferential access to information and investment opportunities. This can also be a weakness, however, since they are also susceptible to local government pressure to support new ventures whose risk and return prospects are not attractive. Also they are not able to attract the most experienced or capable managers, so their ability to assess, monitor and intervene in new venture management is limited.

University VC firms (UVCFs) emerged in large numbers from 2000. They benefit tremendously from their university ties, giving them privileged access to new venture investment opportunities, as well as intimate information about the ventures. On the other hand, they also suffer from some of the same weaknesses as the GVCFs. Specifically, their investment opportunities are in practice limited to those that emerge from the university, and they do not have the managerial expertise related to venture capital investing. Another weakness is that the universities usually are not cash-rich, so they depend more and more on other sources of investment capital; again publicly listed and cash-rich enterprises have become their primary backers. Although there are examples of research institutes founding VC firms (for example, the Chinese Academy of Science is a major investor in Shanghai New Margin Ventures – Shanghai Lianchuan Touzi), they are too few to represent a major category of VC firm. They do, however, have the same advantages and disadvantages as UVCFs.

A wave of corporate VC firms (CVCFs) were founded in response to the 'No. 1 Proposal' of 1998, and they now represent the majority of VC firms operating in China. Beijing High-Tech Venture Capital Ltd and Beijing Venture Capital Ltd were the first CVCFs, founded in October 1998. Their strong government backing, however, causes many to perceive them as firms under the Beijing government's commercial holding company. From early 1999, there was a wave of true corporate-backed CVCFs, although they still sought local government support. Their managers typically come from securities firms, banks or industry.

Listed companies have been the primary source of funds for CVCFs (see

Table 7.7 Comparison of government, corporate, university and foreign venture capital firms in China

	Government VCF	Corporate VCF	University VCF	Foreign VCF
Basic features				
Ownership or legal form	SOE and limited corporation	Limited corporation	Limited corporation	Limited partnership
Initial/primary investor	Local government	Listed companies	University industrial group, other firms	Pension or other funds, corporations
Top manager background	Government bureaucracy and SOE	Securities firm or bank, industry	University's enterprise group or other firms	Foreign VC funds, investment banking
Primary motivation and objectives	Promote local high-tech industry and commercialization	Higher ROI than alternative investments; related business opportunities	Commercialization of university's S&T achievements	High ROI
Investment focus	High-tech	High-tech	High-tech	High growth/potential
Preferred stage of investment	Early	Late, expansion	Early	Growth
Investment time horizon	3–5 years	3–5 years	Not clear	3–7 years
Follow-on investment	No	Varies	Varies	Yes
Geographic distribution of investment	Local	Local, regional	Universities and regional	Major metropolitan areas (Shanghai, Beijing, Guangzhong, Shenzhen, etc)
Internal incentive system	Salary plus bonus	Salary plus bonus	Salary plus bonus	Salary plus carried interest

Strengths	Government base provides ready-made channels to government and access to or information about policies and projects	Strong financial base (proceeds from listing, cash flow from operations) gives them investing flexibility. Industry base gives them management and operational expertise that they can draw on for selecting and monitoring investments, as well as form base for related diversification and pursuing new business opportunities opened up by a new venture	Strong technology base benefiting from R&D activities and concentration of personnel in university; access to primary source of science and technology in China. University link provides them preferential access to those resources	Professional experience in financing and managing start-ups and high-growth firms. Can draw on experience in other markets, link Chinese firms to business partners and markets abroad through foreign network of investees and related business activities. Expertise in decision-making and VC cycle, especially exit decision
Weaknesses	Objectives and incentives are split between financial and social returns to investment; weaker internal incentives than FVC. Investments influenced by policy objectives. Managers may not be familiar with firm management practices, systems, procedures, etc	Short-term investment horizon driven by need (in listed companies) to show annual performance; availability of funds subject to firm's current operational performance in core business. Corporations not experienced with managing high-risk investments represented by investee firms. Weaker internal incentives than FVC	Lack of business management experience. Same problems of internal incentives as GVC and CVC. Investment opportunities limited to those emerging within the university	No strong relationship with major organizations (government, enterprises, universities) in China, so no preferential access to domestic sources of related resources. Must expend time and effort on establishing relationships to gain access to investment opportunities
Future issues	Local governments will not inject additional capital, so short-term investment capacity depends on ability to find alternative sources of investment funds and, later, returns to investments and financing	Role in financing tech-based firms will continue to be important, and corporations will continue to be primary source of funds if pension fund, insurance cannot enter venture capital industry	Lack of key expertise (firm management, VC investment) will drive them closer to corporate VCs and foreign VCs	Further opening of China and their own good performance will allow them to exploit their VC expertise even more in China. Such expertise and linkages outside China continue to make them attractive to potential investee firms

Table 7.8a VC firms: government-backed cases

Government VC Province A high-tech investment	Government VC City S venture capital	Government VC City C venture capital
Foundation		
Est. 6/2000, 100% government-funded with RMB600 million from Shangdong provincial government; as a limited corporation. To promote high-tech in the region, along with economic development; also to 'compete' with the efforts of other provincial and city governments	Founded 1/1998, 100% government-funded, with RMB150 million from Shenyang city government; as a limited corporation To achieve greater returns, promote local S&T industry development and stimulate the local economy.	Founded 1999, 100% government funded, with RMB300 million, as a limited corporation To promote the commercialization of local S&T achievements, provide support and added value services to develop firms
Management and staff		
Initial managers include 15 with MS or PhD degrees, three studied abroad; others from provincial government and Shandong Trust and Investment Corp.	Initial 10 managers all from the local government. Number of investments is large, but small staff	Initial 10 managers from securities firms, investment banks and industry Staff average 5–8 years of investment experience
Funds and portfolios		
17 investments, some related to Shangdong Technology Market, and investment in listed company Funds to be increased to RMB1.2 billion by end of 2002, later to RMB2 billion. Planning a province-wide fund with local cities, and a joint fund with foreign VC firms	NA	Currently 23 investments Plan to increase funds under management, invest in other regions (particularly S&T zones) and help develop the equity market Funds increased to RMB500 million in 2001 and to RMB600 million in 2002
Investment strategy		
Targeted growth and expansion stage, and only in Shangdong	Targeted expansion stage in Shenyang, in new materials and biopharmaceuticals	Targeted early and expansion stage investments in Guangzhou, in IT and biotechnology, and in

Investments concentrated in new materials for electronics, IT, precision processing and traditional medicine Seldom coinvest Would like to restructure into an investment fund and management company, and eventually become a large-scale investment and management holding company focused on high tech and investments nationwide	Primarily evaluate management team, future cash flow and market prospects	areas of medical devices and instruments Joint investments with US and Singaporean companies to establish a venture investment management consulting company and pursue joint investments.
Relationship with investees		
Involvement in investees through board membership, financial audit, visits, management consulting	Involvement through board membership and provision of management consulting	Involvement through board membership and provision of management consulting
Learning and networking		
Send managers to training, and host outside trainers and consultants. Involvement in Shandong Technology Market and cooperation with city governments for other investment funds; participate in national and foreign cooperative investment networks. Signed cooperative agreements with accounting and law firms; searching for outside intermediaries and partners to improve investment activity and management	Learn by doing Use links to government bureaucracies, shareholders	Participate in study meetings, exchange associations, etc, sponsor training, and established joint venture investment consulting company. Founded venture capital promotion association, Guangzhou Industrial Rights Market, S&T guarantee and Guangzhou International Technology Park Cooperative relationships with banks, brokerages, universities, research institutes, investment organizations, intermediaries and incubators
Performance measures		
No exits so far Reputation and impact limited to Shangdong	Five exits: one listing on the main board, four by acquisition Not well known outside Shenyang	No exits so far Well known in Guangzhou

Table 7.8b VC firms: corporate-backed cases

Corporate VC City B venture capital	Corporate VC City Z venture capital	Corporate VC Legend capital
Foundation		
Est. 10/1998 with RMB500 million from six Beijing-based holding companies (utilities, investments, services, etc) as a joint stock company	Est.1999 with RMB700 million by Shenzhen government's holding company and listed companies; as limited liability company To increase competitiveness of Shenzhen's S&T industry vis-à-vis other regions	Est. 3/2001 with RMB200 million, mainly from Legend Holdings and its employees as a limited liability company; single-company venture capital arm. Also manage HK$50 million fund
Management and staff		
Managers from investor firms, with govt. and firm experience and affiliations Currently 39 staff, 35% with PhD or MS, 18% with foreign study experience	Top managers from major security firms 48 staff members, including 27 with MS, eight with PhDs	Four senior managing directors, 10 investment project managers and additional senior support staff
Funds and portfolios		
Currently 43 investments	Funds increased to RMB1.6 billion in 2000; eight different funds: three specialized funds, three Sino–foreign joint funds; seven regional investment management companies. 61 direct investments (RMB530 million) and 16 investments in other funds (RMB320 million)	More than seven investments; plan to establish a second fund
Investment strategy		
Investments in Beijing (with an exception in Xian) in IT, biopharmaceuticals, new materials and modern agriculture Target 30% seed capital, 60% growth/expansion, and 10% other activities, subject to adjustment	Most investment in Shenzhen, but also in Shanghai, Chengdu, Wuhan, Harbin, etc Lead investor in approximately half of investments; coinvestment cases increasing. Established Shenzhen Venture Investment	Concentrate in IT, esp. communications and network equipment, corporate software, IT services, semiconductor fabrication Prefer to invest in start-up and early expansion stages, some seed and pre-IPO

Working to increase investment scale and pay more attention to exit Increasing coinvestment Government background makes some see company as having some characteristics of traditional SOE	Group in 10/2002 to capitalize on trends of internationalization, regionalization and specialization	Undertakes each stage of due diligence Investment horizon 5–7 years
Relationship with investees		
Limited involvement in investees, but do help to establish management systems, linkages with relevant bureaucracies and tap other funds (esp. SME Innovation Fund)	Involvment in investee boards and post-investment monitoring Consult on commercial scale, follow-on financing, risk management and market development	Only intervenes as board member (depending on the share percentage) or in operations if investee requests Legend's help, although their involvement is greater than most other Chinese VC firms'; for example, to introduce suitable suppliers or customers or address management challenges
Learning and networking		
Cooperate with Tianjin VC firm; promote internal investment expertise as a basis for cooperating with foreign investment funds Receives some support from National Development Bank to further venture capital industry and policy; cooperation with China Construction Bank to develop new technology ventures; cooperative agreements with Beijing 10 strongest incubators, etc	Training, strategic change, research and joint development of venture investment companies with organizations outside Shenzhen. Participate in technology markets, cooperate with financial organizations, form Sino–foreign joint venture investment bank; other linkages with foreign venture capital; cooperate with leading corporate groups in particular industries	Draws on Legend's sources of technical and managerial expertise, links to industry actors
Performance measures		
No listing of investee companies, but three sales of partial stakes Strong position in Beijing	Largest VC in China, major impact in VC industry	Currently the most representative corporate VC in China, but still needs to develop expertise and experience

Table 7.8c VC firms: university and foreign-backed cases

University VC Univ.T science park investment	Foreign VC IDG venture capital	Foreign VC ChinaVest
Foundation		
Est. 11/2000 with RMB50 million as a limited liability corporation with three investors (Tsinghua's Science Park Development Center, and two organizations in other regions) To meet the financing demands of start-ups in Tsinghua, and exploit opportunities created by Tsinghua's S&T capabilities and outputs	Parent organization established in USA as subsidiary of IDG publishing; limited partnership. Created US$50 million region-focused Pacific Development Venture Fund in 1993; investments in 15 Asian countries. First entered China in 1994 with joint venture funds with S&T commissions of Beijing, Shanghai and Guangdong	Est. in 1983 by partners from the USA To take advantage of human resources and new opportunities from China's reforms, but also throughout Greater China
Management and staff		
Most employees have MS or PhDs; managers have overseas study and work experience; investment managers have MBAs from Tsinghua, but limited investment experience	Managers and partners are primarily PRC nationals with an international background HQ for China in Beijing, with branches in Shanghai, Guangzhou, Tianjin, Shenzhen, in addition to US offices in Boston and Silicon Valley	Partners from the USA, although each have more than 20 years of experience in Asia in investment banking and management consulting. Offices in Beijing, Hong Kong, San Francisco, Shanghai and Taipei
Funds and portfolios		
Seven investments	100% foreign funded In 1998 established a fund of US$1 billion for investment in China 85 investments in China	Five funds under management, totaling US$300 million Over 30 investments in Greater China
Investment strategy		
Main investment focus is IT-related ventures Non-controlling shareholder Relatively many coinvestments; usually not the	Investments concentrated in high-tech industries, esp. global networks, information services, software, telecommunications, networking	Regional focus, but not an industry focus (high-growth, not just high-tech) Original focus on manufacturing, now services

lead investor	Choose investments in ventures with strong prospects of providing products or services to growing markets	and high-tech since end of 1990s; diversifying into name brand consumer services, media, telecom and IT, distribution, and value-added manufacturing Usually lead investor
Relationship with investees		
Involvement through participation on boards, financial audits, occasional visits	No involvement in daily operations Assist investees in strategy, financial planning, sales and distribution network, and market development; bring in outside consultants, introduce potential partners	Help investees attract managers and capable board members, develop operations; develop M&A opportunities
Learning and networking		
Established internal management system for project management, investment management, human resource management, etc; emphasize training, domestic and international conferences, encourage self-study Tsinghua connection gives it access to research institutes nationwide Cooperative agreements with a law office and two accounting firms	Draws on IDG's global network for resources	Board members include public officials Continuous study of Greater China markets, government policy, and developments; also attract top Chinese human resources Act as a bridge between US capital and Chinese entrepreneurs
Performance measures		
Still developing, and no strong reputation or major impact on VC industry	10 exits (eight by acquisitions, two by overseas listing – NASDAQ and Hong Kong GEM) One of the most influential VCs in China By becoming involved in China early, has had an influence on the course of the industry	Exits include two listings on NASDAQ and numerous acquisitions Wide reputation inside and outside China as first and successful China-specialized fund; has attracted attention domestically and abroad

Table 7.9 for examples). These firms had received massive infusions of cash when they went public, much more than they could use on productive internal investments. They were also looking for promising new areas of business. Many thought that the Chinese government would soon establish a second board for listing new ventures (on the model of Hong Kong's Growth Enterprise Market) and that this would provide a lucrative exit for venture-stage investors. The media had also fueled popular interest in venture capital, although the coverage was shallow and did not educate the public about the inherent risk in such investments.

Table 7.9 Examples of listed company involvement in VC firms

Neusoft (Dong Da A Er Pai Ruan Jian Gong Si), a leading software firm, has invested approximately 70% of Liaoning East Information Industry Venture Capital's RMB100 million registered capital. That VC firm focuses on IT, especially software and digital technology projects
Wanxiang Group (Wan Xiang Ji Tuan), an automobile parts manufacturer, invested RMB200 million of the RMB300 million registered capital of Wanxiang Venture Capital Co., Ltd, which funds biotech and pharmaceutical, IT, environmental protection, new materials and other technology-based projects
Ancai High-Tech Co. (An Cai Gao Ke), a producer of key components for TVs, invested RMB200 million, representing 97% of a new venture capital firm's registered capital
China Youth Travel Service (CYTS, *Zhongqinglu Konggu*), a diversified firm in the travel industry, is a major shareholder of Beijing Venture Capital Corp., having invested RMB125 million
Beijing International Trust and Investment Corporation (*Beijing Guoji Xintuo Touzi*), a diversified investment company, is also a major investor in Beijing Venture Capital Corp. (RMB125 million), as well as in Tianjin's Taida Venture Capital Corp
Capital Iron and Steel (*Shougang*), one of China's largest integrated steel manufacturers, is a major shareholder of both Beijing High-Tech Venture Capital Corp. and Tsinghua Venture Capital Co
Nearly half of Shenzhen Venture Capital's RMB1.6 billion registered capital was from listed firms
Of Tsinghua Unisplendor Venture Capital's 12 shareholders, 11 are listed companies

As a result, by the end of 2001, 132 public listed companies had invested in CVCFs, accounting for 11 per cent of all listed companies. Additional sources of backing for CVCFs include unlisted firms with large cash flows, individual investors and foreign firms. Often these investors are directly involved in the industries in which the CVCF invests, and are able to draw on their backers' industry and managerial capabilities to assist ventures in which they invest. In addition to any financial returns to their investment, the CVCF can help the corporate investors identify related new business opportunities. At the same time, new ventures benefit from these links to potential suppliers and customers, in addition to the financing that they receive.

CVCFs and their backers, however, had invested in ventures with the expectation that the investees would list quickly. They have not proved themselves to be interested in long-term development of the new ventures. As the government has postponed establishing a second board, however, their timeframe for realizing a return on their investments is becoming unexpectedly longer and longer. A number of these CVCFs have suffered heavy losses, pushing some into bankruptcy.

Finally, foreign VC firms (FVCFs) have entered China and become a major source of new venture financing. By 2001, eight of the top 10 VC investors in China were foreign firms, and 14 of the top 20 (Table 7.10). Like the domestic CVCFs, most of the FVCFs are backed by multiple investors, although a few (for example, Intel Capital) are the investment arms of single firms.

Several characteristics distinguish FVCFs from the other types of venture capital firms operating in China, besides their legal form (usually limited partnerships). One fundamental difference is their focus on high-growth or high-potential investment targets, not necessarily high-tech. They also have greater expertise in venture capital management. Zhang and Jiang (2002) found that the managers in domestic venture capital firms averaged 2.1 years of relevant experience, while those of FVCFs operating in China averaged 11.9 years. These firms also have stronger incentives that both retain managers and encourage them to manage investments for longer-term gains. The FVCFs are usually also able to provide linkages to potential customers and partners in foreign markets.

New Technology Ventures

The focus of all of the government and firm actors introduced so far is, of course, China's new technology ventures. Although this category includes a wide variety of firms, there are a number of characteristics that they commonly share. First, a large number of ventures are spin-offs from research institutes or universities. Until recently, these were the only sources of entrepreneurs. Increasing numbers, however, are now coming from industrial firms. Linkages to such organizations, and their tacit or explicit support, were a critical feature of China's earliest technology-based ventures founded

*Table 7.10 VCF ranking by cumulative investment in China**

Ranking	VC firm
1	IDG VC
2	Chinavest Ltd
3	Intel Capital
4	Shenzhen Venture Capital Co. Ltd
5	H&Q Asia Pacific
6	WI Harper Group
7	Baring Private Equity Partners (Asia)
8	Goldman Sachs (Asia) Ltd
9	Vertex Management
10	Walden International
11	Beijing Venture Capital Co. Ltd
12	Guangdong Technology Venture Capital Group (GVCGC)
13	Shanghai Venture Capital Co. Ltd
14	NewMargin Venture Capital
15	Draper Fisher Jurvetson
16	Warburg Pincus
17	Government of Singapore Investment Corporation
18	Soft Bank China Venture Capital
19	The Carlyle Group
20	Transpac Capital

Note: * As of 2001.

Source: http://www.zero2ipo.com.cn.

in the mid-late 1980s: Stone, Founder and Legend (Lu, 2000). They continue to be necessary in practice today, even if not a formal requirement. These spin-offs range from small numbers of individuals, to sub units within the source organization, to entire organizations in the case of corporatized institutes. The technology and capabilities that form the basis of the new venture are embodied in these individuals, sub units and organizations. Individuals without linkages to such organizations, in contrast, often face an uphill battle in securing funding; in practice, they do not have access to most sources of venture capital unless they have special ties to industrial firms.

Another feature is the industries in which these new ventures, and most venture capital financing, are concentrated. They are insignificant in traditional industrial sectors, although FVCFs are more likely to invest in these areas, and more and more domestic VC firms are seeing such sectors as promising. Still, the majority of new ventures are in the information technology industry: hardware, software and services. Because IT has an impact on all sectors, however, the new technology developed and diffused by these ventures has a broad impact on the economy as it increases the efficiency and effectiveness of enterprises in all areas. More recently, biotechnology-related ventures have emerged as a significant area, a recognition of the potential this area represents as well as a shift in investor attention since 2001 from disappointing Internet projects. MoST now cites a somewhat expanded list of technology areas as priorities for development and commercialization: electronics and IT; biotechnology; new materials; integration of optical, mechanical and electronic components; new energy, high-efficiency energy, energy-saving technology; and environmental protection (Chen, 2002).

VC Firm Involvement in New Ventures

A comparison of domestic and foreign venture capital firms vis-à-vis their investees in China provides insights not only into the nature of the relationship between VC firms and investees, but also into the stage of development of China's VC firms themselves. Zhang and Jiang (2002) find a number of key differences, supporting those uncovered by Bruton and Ahlstrom (2002) in a recent study.

First, Chinese VCs are less active in their monitoring of investee management than foreign VCs. For example, foreign venture capital firms require financial reports more frequently. While almost all FVCFs require monthly financial reports, only two-thirds of domestic VCs required monthly reports. FVCFs also retain veto rights, while fewer domestic VCs obtain such rights.

Second, domestic VCs exercise weaker influence over their investee management decisions than do FVCFs. For example, they use staged investment in the same round of financing less frequently than FVCFs. Also they are less likely to make follow-on investment and cash flow rights of entrepreneurs contingent on the venture's performance. While domestic VCs are beginning to introduce stock option plans more generally into investee firms and often only to top management, FVCFs almost always introduce stock options into investee firms and for all employees.

Finally, the domestic VCs provide much less to investees in terms of value-added services. While FVCFs usually take part in board meetings at least once a quarter (and often monthly), less than half of the domestic VCs participate so frequently. Indeed, an underlying difference between these types of firms is that the domestic firms in general do not see addressing operational issues of investees as an important part of their development, or their role as investors. Instead, they concentrate their monitoring and participation on the financial aspects of the investee firms.

One reason for some of these differences, already alluded to in the description of FVCFs and domestic venture capital firms, is that domestic VCs are much less experienced than their foreign counterparts. This can explain their more restrained involvement in investee firms: they do not have the experience base to justify taking a leading role in many top management issues. It also explains the limited value-added services they provide to investees. Although capital has been raised quite quickly, the experience and expertise to invest, monitor and support investee firms takes much longer to develop.

FVCFs also tend to invest at earlier stages than domestic VC firms. Since mid-2001, as new investment funds became scarcer and domestic VC firms needed to generate profits, this divergence has become even more pronounced. VC in China has shifted from the development stage to latter stages, such as the growth and pre-IPO stage firms. Of course the government policies of each region differ, and the investment strategies of VCs in different regions differ, in terms of timing and so on. For example, VC investments in Beijing are concentrated on post-development stage ventures, while in Shanghai start-up stage investments are more common. Similarly, VCs backed by universities tend to invest in start-up stage ventures. The CVCFs, however, are more focused on realizing returns sooner rather than later, and are increasingly wary of inherently risky and uncertain projects. Since they are the primary source of venture funds, this shift represents a contradiction between the desires of the government for VC to nurture early-stage high-tech firms and the logic of the market represented by VC firms' decisions.

Domestic and foreign venture capital firms, however, face the same challenge in interacting with management in new ventures. Although this is perhaps not limited to China, many local entrepreneurs are extremely reluctant to allow 'outsiders' (including investors) into the firm. They tend to perceive such outside involvement as a potential loss of control or power. This perception has been exacerbated by the media, which have tended to position 'capital' and 'knowledge' (venture capitalists and entrepreneurs, respectively) as opponents, rather than as working toward a common goal and mutual gain.

CONCLUSIONS

Venture capital and the commercialization of new technology has emerged as a major contributor to China's development and sustained economic vitality. The central and local levels of government have played a key role in the development of both the venture capital industry and the new technology ventures it has supported. Rather than a systematically executed plan, however, the current state is better conceived as the result of an emerging process, the outcome of a coevolution of bureaucratic structures (incubators, technology zones), policies, institutions and new types of firms.

China's venture capital system is still immature in terms of the resources and capabilities of most of the constituent organizational actors, as well as the institutional environment in which they operate. Therefore, while the indicators of the system's ability to support the development of new technology ventures and the commercialization of China's S&T resources are impressive, there are number of major weaknesses. Currently, venture capital firms do not have the expertise or operational mechanisms to select and manage new technology ventures adequately, nor have they been able to add much value beyond financing. Because their incentive structure biases them towards late-stage investment projects, these venture capital firms are not acting as a channel of funds to true start-ups, in contrast to the government's hopes in promoting venture capital. After an initial spike of activity in the late 1990s to 2001, the system is now in a stage of consolidation and evolutionary change. This period should allow domestic venture capital firms to build up necessary experience and expertise, and also allow government actors to introduce appropriate policies, regulations and incentives to support the venture capital industry so that it has a positive impact on new technology venture development.

NOTES

1. In this chapter, we use the term 'institutions' in the sense of North (1990), as 'rules of the game' and including disembodied systems of practices, norms and regulations; the legal system and financial system are in this sense institutions; in contrast to Nelson (1993) and others who use the term to indicate actors (such as universities) or a cluster of actors (educational system).
2. For a fuller account of this transition in China's national innovation system, see Gu (1999), Liu and White (2001).
3. The main channel for foreign private equity in China before 1992 was China Direct Investment Funds; see Bruton and Ahlstrom (2002).

REFERENCES

Bruton, Garry and David Ahlstrom (2002), 'An institutional view of China's venture capital industry: Explaining differences between China and the West', *Journal of Business Venturing*, **17**, 1–27.

Chen, Mingxuan (2002), 'Torch program and development of small and medium-sized technology-based enterprises in China', presentation at The Second

International Training Workshop on Technological Innovation for Small and Medium-sized Enterprises Based on Science and Technology, Beijing, September.

Child, John (1994), *Management in China during the Age of Reform*, Cambridge: Cambridge University Press.

Cohen, W. and D. Levinthal (1990), 'Absorptive capacity: A new perspective on learning and innovation', *Administrative Science Quarterly*, **35**, 128–52.

Gao, J. and J. Fu (1996), Qiye jishu chuangxin de guanjian wenti' (The key problems of technological innovation in business firms), *Zhongwai Keji Zhengce Yu Guanli* (*Science and Technology International*), **1**, 24–33.

Gu, S. (1999), *China's Industrial Technology: Market Reform and Organizational Change*, London: Routledge.

Lu, Qiwen (2000), *China's Leap into the Information Age: Innovation and Organisation in the Computer Industry*, Oxford: Oxford University Press.

Liu, Xielin and Steven White (2001), 'Comparing innovation systems: A framework and application to China's transitional context', *Research Policy*, **30**, 1091–114.

McDonald, T. David (1990), *The Technological Transformation of China*, Washington, DC: National Defense University Press.

Naughton, B. (1994), *Growing out of the Plan: China's Economic Reform, 1978–92*, Cambridge: Cambridge University Press.

Nelson, R. (ed.) (1993), *National Systems of Innovation: A Comparative Study*, Oxford: Oxford University Press.

North, Douglass (1990), *Institutions, Institutional Change and Economic Performance*, Cambridge: Cambridge University Press.

Steinfeld, Edward. 1998. *Forging Reform in China: The Fate of State-Owned Industry*. Cambridge: Cambridge University Press.

White, Steven (2000), 'Competition, capabilities and the make, buy or ally decisions of Chinese state-owned firms', *Academy of Management Journal*, **43**, 324–41.

White, Steven and Xielin Liu, (1998), 'Organizational processes to meet new performance criteria: Chinese pharmaceutical firms in transition', *Research Policy*, **27**, 369–83.

White, Steven and Xielin Liu (2001), 'Alternative transition trajectories for market structure and firm strategy in China', *Journal of Management Studies*, **38**, 103–24.

Zhang, Wei and Yanfu Jiang (2002), 'The relationship between venture capitalists' experience and their involvement in the VC-backed companies', paper presented at the Global Finance Conference, Guanghua School of Management, 27–29 May.

Zucker, Lynn (1986), 'Production of trust: Institutional sources of economic structure, 1840–1920', *Research in Organizational Behavior*, **8**, 53–111.

8. Venture capital and innovation: the Indian experience

B. Bowonder and Sunil Mani

INTRODUCTION

The purpose of the chapter is to understand the value-adding support functions which venture capital companies, render to their investee companies, especially in the Indian context. As seen in the earlier chapter by Mani and Bartzokas, the VC industry in India has exhibited significant growth during the 1990s. In fact one sees a coevolution between the growth of VC and the growth of the IT industry in the country. This chapter is structured into three broad sections. The first section traces the growth of the VC industry in the country, while in the second section we discuss detailed cases of 11 investee firms. Finally in the last section, we summarize the main insights obtained from these case studies.

THE INDIAN VC INDUSTRY

The government took a major initiative in establishing a VC industry in India towards the latter half of the 1980s. Its genesis and subsequent growth up to the mid-1990s are discussed in Mani (1997). The initial growth of the industry was regulated by the VC guidelines of 1988 which, as noted in Mani (1994), were an important document in the sense that it clearly laid out the scope of venture financing in the country in terms of specifying (a) stage of financing, (b) instrument of financing, and (c) industry of financing. The ideal type of venture financing envisaged was equity financing at the early stage of technology-based enterprises. There are currently three[1] separate government regulatory policies, namely the Securities and Exchange Board of India, SEBI (VC funds) Regulations of 1996, the Guidelines for Overseas Venture Capital Investments issued by the Ministry of Finance (1995) and the Central Board of Direct Tax Guidelines for Venture Capital Companies (1995), which govern the functioning of VC funds in the country. Of the three, the SEBI regulations are

Table 8.1 Main provisions of the venture capital fund regulations of 1996

Investment conditions	Scope
Minimum investment in a VC fund	A VC fund[1] may raise money from any investor, whether Indian, foreign or non-resident Each scheme launched or fund set up by a VC fund shall have a firm commitment from the investors for contribution of at least Re50 million
Investments conditions and restrictions	A VC fund shall not invest more than 25 per cent of the fund in one undertaking At least 75 per cent of the investible funds shall be funded in unlisted equity shares or equity-linked instruments Not more than 25 per cent of the investible/funds may be invested (a) by way of subscription to an initial public offer of a VC undertaking[2] whose shares are to be listed subject to a lock-in period of one year; (b) by way of debt or debt instruments of a VC undertaking in which the VC fund has already made an investment by way of equity
Lock-in period and listing on any recognized stock exchange	No VC fund shall be entitled to get its units listed on any recognized stock exchange until the expiry of three years from the date of the issuance of units by the VC fund

Notes:

1. A VC fund is a fund established in the form of a trust or a company including a corporate body and registered under the regulations and which (a) has a dedicated pool of capital raised in a manner specified in the regulations and (b) invests in a VC undertaking in accordance with the regulations.
2. A VC undertaking is a domestic company whose shares are not listed on a recognized stock exchange in India and which is engaged in the business of providing services, production or manufacture of articles other than (a) real estate, (b) non-banking financial services, (c) gold financing or (d) any other activity which is not permitted under the industrial policy of the government of India.

Source: Securities and Exchange Board of India, http://www.sebi.com/.

the most important. An examination of the chief provisions of these regulations shows that a number of important issues related to what makes VC investment distinct from other forms of investment are missing (Table 8.1).

There are three important dimensions, that merit our attention: (a) the VC investment must be in manufacturing activity, (b) the investment must be largely in the form of equity investments, and (c) the VC fund must hold on to the shares in a VC undertaking for at least three years. However the regulations do not make a finer distinction in manufacturing or services between technology oriented and otherwise and they do not specify the stage of financing. The 'value-added support' a VC fund can give to a VC undertaking is also unspecified and the provision of this support is an important distinguishing aspect of venture capital financing.

Some Stylized Facts about the Growth of the Venture Capital Industry in India

The new VC investments, have on an average, increased by 81 per cent per annum during the period 1992–2001 (Figure 8.1). Significant increases, however, have happened only since 1999 and this is due to the launching of a state initiated VC funds specifically for the IT industry (see Chapter 6 in this volume). Despite this phenomenal increase, VC investments are still less than 1 per cent of gross fixed capital formation. Even though the total VC pool (funds under management) has increased significantly over the past decade (Table 8.2), it is still much smaller compared to other newly industrializing countries of Asia. For instance the ratio (in 2000) of the total pool of VC funds in India to those in Hong Kong, Singapore, Korea and Taiwan works out at 0.10, 0.31, 0.48 and 0.49, respectively. The first significant decrease in the rate of growth of total capital under management

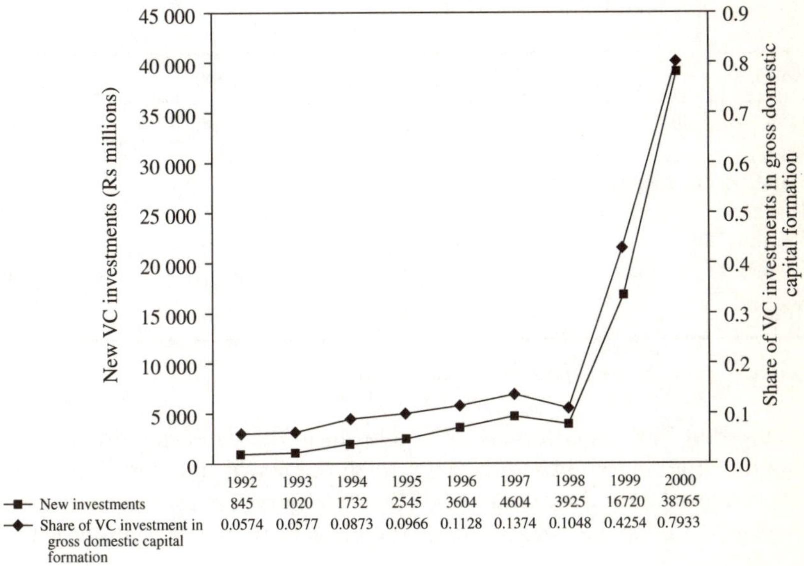

Source: *Asian Venture Capital Journal* (2002), Government of India (2002).

Figure 8.1 Trends in new VC investments in India, 1992–2000

Table 8.2 Dimensions of the VC industry in India, 1991–2001 (Rs, millions)

	Total funds under management	New investments made during the year	Total capital raised during the year	Divestments
1991	2385		590	
1992	3276	845	891	
1993	4646	1020	402	
1994	7638	1732	2824	
1995	10063	2545	1822	
1996	28113	3604	15516	
1997	41668	4604	10800	
1998	44720	3925	2530	89
1999	79452	16720	11047	279
2000	135053	58765	54822	1290
2001	117830	54124	19532	14750

Source: *Asian Venture Capital Journal* (various issues).

Table 8.3 Source of funds for VCs in India, 1999–2001 (percentage shares)

Source	1999	2000	2001
Corporations	61	48	46
Banks	15	21	23
Government agencies	9	5	5
Insurance companies	7	8	10
Pension funds	2	10	12
Private individuals	6	8	4
Total	100	100	100

Source: *Asian Venture Capital Journal* (various issues).

occurred in 2001. This decrease is largely due to the significant divestments that occurred in that year, especially when the new investments had actually increased during the same year. Thus the decrease in total capital under management may be a reflection of the busts in the IT/Telecom boom of 2000.

Corporations are the major source of funds to the Indian VCs, although their share has tended to come down (Table 8.3). Though the explicit role of government as a source of funds is low, one should be careful in inter-

Table 8.4 Distribution of VC disbursements in India, by stage, 1999–2001 (percentage shares)

Stage	1999	2000	2001
Seed	7	9	7
Start-up	47	40	36
Expansion	36	42	49
Others	10	9	8
Total	100	100	100

Source: *Asian Venture Capital Journal* (various issues).

Table 8.5 VC disbursements in India, by technology, 1999–2001 (percentage share)

Technology	1999	2000	2001
Computer-related	16.4	19.0	16.8
Electronics	4.5	4.4	4.7
Information technology	11.0	18.1	16.7
Manufacturing – heavy	5.6	3.4	2.6
Medical	7.7	6.0	4.7
Telecommuncations	10.9	11.0	22.6
Total technology-based ventures	56.1	61.9	68.1

Source: *Asian Venture Capital Journal* (various issues).

preting it so because most of the banks (in this case development banks such as ICICI and IDBI) too are owned by the government. Approximately 70 per cent of the funds have emanated from abroad, and the share of Asian countries has shown a considerable increase.

Some 50 per cent of the VC disbursements in the country (Table 8.4) have gone towards the early stage. India is very unusual in this respect, as except for Israel, in most countries, including the USA, VC disbursements are largely sought at the expansion stage.

Technology-intensive sectors such as computers and information technology[2] account for the single largest share. Again, some 60 per cent of the disbursements have gone towards technology-based ventures (Table 8.5). Thus the investment pattern of the Indian VC industry presents an idealistic picture, assisting, by and large, technology-based ventures at their early stage.

The majority Indian VC firms have their headquarters in Mumbai, which is the financial capital of the country, but it is interesting to note that about 20 per cent of them are based in Bangalore, the south Indian city which is often referred to as the 'Silicon Valley' of India.

THE CASE STUDIES OF VC-ASSISTED FIRMS

A number of VC-assisted firms were visited by one of the authors. The following are the firms covered in the study. After interviewing the chief executives of 26 firms, 11 firms were selected and analysed. The details of the cases are first mapped out and subsequently the insights to be drawn from them are discussed.

Tejas Networks India

The vision of Tejas Networks is to create state-of-the-art products and solutions in the telecommunications and optical networking arena. Tejas Networks was founded in 2000. The firm developed software-differentiated optical networking products that provide high price/performance in their class, enabling carriers to maximize revenue-generating services while optimizing their overall network costs. Tejas Networks also partners with leading third party equipment vendors to build intelligent optical networks for its customers. During 2002–3 Tejas won 10 new customers against global competition, together with large orders from many Indian and foreign firms. During the year it grew by 180 per cent in revenue terms.

Founders and their experience Sanjay Nayak is the co-founder and chief executive officer. He worked as the managing director of Synopsys India and also he had experience in working with Viewlogic Systems and Cadence Design Systems, in the USA. Dr K.N. Sivarajan is the co-founder and chief technology officer of Tejas. He was a professor at the Indian Institute of Science, having worked prior to this in the IBM Watson Research centre. He received his PhD from California Institute of Technology. Arnob Roy is the third co-founder and earlier he worked with Synopsys India. The Tejas team consists of outstanding professionals with a wealth of experience in deploying carrier class[3] optical networks in India and the USA.

Origin of the idea Mr Nayak and Dr Sivarajan decided to create something new for self-actualization. They wanted to create a world-class product company, as they wanted India to develop innovative telecom products. Mostly, Indian firms were in software services. They wanted to create products from India. This urge made them seek venture capital as they had innovative ideas. Tejas is pioneering the trend for Indian companies moving up the value chain and developing state-of-the-art products using trends in software and hardware design.

Venture investors There were three venture investors for Tejas Networks in the first round: Mr Gururaj Deshpande, Chairman of Sycamore; Sycamore Networks, a publicly held corporation; and ASG Omni LLC, a financial agency. In the first round the three investors provided US$5 million. In the second round Mr Deshpande, Intel Capital and ILFS invested US$6.7 million. Intel Capital is the strategic investment arm of Intel.

Products The main products of Tejas are cost effective SDH Multiplexer equipments designed to manage bandwidth and derive services from the optical core to access. Innovation in optical networking requires high levels of software and hardware integration capabilities. Tejas has undertaken the design and deployment of optical networks. Through innovation and learning, the firm is able to compete with global firms like CISCO, Nortel and Lucent. Tejas combines the cost advantage of India and the innovative strength of its founders. The optical products are based on the dense wave diversion multiplexing and optical amplification to transmit data optically at aggregate rates exceeding one terabyte per second over distances of a few thousand kilometres on a single strand of fibre.

Tejas Networks India Ltd, an optical networking start-up, launched its intelligent optical access product in India less than a year after its start. Intel Capital announced funding after the product was announced. The nine-month-old company immediately won its first customer, Tata Power to deploy the TJ-100 access product. This is the first intelligent optical network in India. The system leverages the capacity creation of DWDM technology and innovative networking software. With the Internet infrastructure market growing at about 20 per cent per annum, Tejas Networks hopes to market its TJ 100 family of products in the global market. As regards venture funding and value addition, Tejas Networks is a knowledge integrator. The firm essentially develops network software and markets Sycamore's optical networking products in India and the Asia–Pacific. It also develops some regionally specific networking products. The venture capital firms supported Tejas in a number of ways:

- the name of Deshpande added reputation and acted as a non-traded externality to attracted VCFs;
- Intel capital helped in assessing the business plans;
- ILFS helped in co-funding through its private equity arm; and
- ASG-Omni helped in developing business contacts.

Strand Genomics

Strand Genomics is a bioinformatics company that develops innovative algorithms and solutions in the field of bioinformatics. Strand's vision is to accelerate the drug discovery process by developing a suite of products for genomics, proteomics and silico-biology. Use of the state-of-the-art knowledge management solutions allows nuggets of knowledge to be extracted from a large pool of data generated by high throughput technology. Though it is a new firm, it has been able to obtain contract research from many global firms. It is likely to receive second-stage funding. The exact amount has not been announced.

Founders A group of scientists and engineers from the USA and India came together to become world leaders in bioinformatics. The founders were computer scientists with complementary skills in clustering techniques, graphics and visualizations, and stringology. All the board members have PhDs and rich domain experience. Dr Vijay Chandru, who is a Professor of Biochemistry at the Indian Institute of Science, came from MIT. The objective of setting up Strand was to develop tools that leverage unique high-end computational skills.

Venture funding UTI Venture Funds picked up a 17.5 per cent stake for an undisclosed sum. UTI Venture Funds picked up a further 17.15 per cent stake in Strand after a thorough assessment. The second stage funding is by Westbridge, an off-shore fund.

Product Strand Genomics has launched two products; 'Soochika' is a micro-array knowledge management tool and 'Sphatika' is an image classification software. The objective is to provide a tool box that addresses the most common problems faced by drug discovery scientists. The company has a total solutions approach to drug discovery. The tools cover modules for

- visualization,
- high-dimensional data analysis,
- micro-array analysis,
- intelligent drug prediction,
- protein modeling, and
- sequence modeling and analysis tools.

Strategy Within a year of its establishment the firm introduced a series of products. Strand's business model is a combination of providing high-end services and building up a suite of products called 'Oyster' to improve the productivity of the drug discovery process. Strand uses a service model that provides revenue on a continuous basis. For example, Strand entered into a partnership with Gladstone Institutes to analyze complex micro-array data. Strand will use its proprietary data analysis techniques to analyze micro-array data sets generated at Gladstone Institute from experiments using Alzheimer's disease-related mouse models to identify certain genes and associated regulatory networks. Strand also entered into partnership with Automated Cell, which is a disease phenotype-driven drug discovery company. Strand provides advanced algorithmic skill sets and software engineering skills to develop products and solutions for Automated Cell's drug discovery platform which quantifies in-vitro disease phenotypes for target prioritization and validation and optimization of drug discovery levels in oncology and immune disease. Recently Strand licensed its micro-array and datamining product to the Bioprocessing Technology Insitute, Singapore. It also entered into a collaborative arrangement with the University of California, San Francisco.

Strand is a unique company with skill sets not normally available. The senior team consists of a group of scientists and problem solvers encompassing the areas of computer science and biology with the requisite skills for drug discovery. Strand focuses on solutions that have resulted in huge improvements in both productivity and interpreting knowledge from genomic data. The solutions that Strand provides are cost-effective and scalable and hence an unbeatable combination. Strand Genomics co-founder Dr R. Hariharan is in the Technology Review TR100 list of 2002. The service-oriented and long-term partnership relationships make Strand's model a fast growth and low risk model. The second-stage funding was announced recently. Strand Genomics and Clingene International form a joint venture to use the data-mining tool, Soochika, to discover patterns that can give insights into diseases. VC support helped in evolving the product and developing markets.

Avesthagen

Avesthagen is a fully integrated biotechnology and bioinformatics company set up primarily to promote research and development services worldwide, making use of proven latest high-throughput technologies and supported by a well trained research team. The vision of the company is to improve the productivity in agriculture and develop agrotechnologies that would lead to value addition in food and pharma products. Avesthagen focuses on contract research for global firms and it is a cost effective research firm in genomics.

Founder The company was founded by Dr Villoo Morawala Patel. He was awarded a PhD in 1993 in plant molecular biology from the University of Louis Pasteur, France and had work experience at the University of Ghent, Belgium. She founded Avestha Gengzaine Technologies in April 1998.

Origin of idea Upon her return to India in April 1998, Dr Patel founded Avesthagen with four employees, using the technology developed by Tata Institute of Fundamental Research (TIFR). Avesthagen raised US$2 million as venture funding from ICICI Ventures, Global Trust Bank and Tata Industries Ltd. The dream of Dr Patel was to invent edible vaccines and new plants using genomics.

Venture capitalists The three institutions that funded the first round (US$1.5 million) are: ICICI Venture Funds, Global Trust Bank and Tata Industries Ltd. ICICI is one of the foremost investors and stakeholders in Avesthagen. GTB offered a loan, which was later converted into Avesthagen equity. Tata Industries picked up a stake in Avesthagen. Avesthagen has engaged Kotak Mahindra and KPMG as investment bankers to facilitate the process of raising the second-round funding. Avesthagen is looking for a funding of $10 million in the second round. In 2001–2, the firm had an income of US$1.5 million and it hopes to break even in 2003 and reach a revenue of US$10 million in five years. It is also a major player in contract research.

Products and services Avesthagen focuses on both products and services. This business model is basically more robust, as services provide for a regular base revenue. Avesthagen essentially provides four services:

- providing user-friendly database application and management for life science companies;
- providing new tools that allow the prediction of complex sequences at the gene and protein level, using customized algorithms and annotation tools;
- providing 3D fold structural insights to protein modeling;
- providing clean vital data from a given bulk sequence.

Avesthagen has developed complementary DNA libraries in three modules, namely standard cDNA libraries, normalized and subtractive cDNA libraries.

Avesthagen was recently awarded a US patent on a segment of rice DNA sequence. This will help in enhancing rice productivity. The second thrust area is 'edible vaccines'. The vaccines will be made part of the gene in a plant food product so that it can be administered easily and in a cost-effective manner. This firm is one of the VC-assisted firms that is focusing on creation of intellectual property. In this case, venture firms helped in assessing the business model for its robustness.

Ittiam Systems

Ittiam is positioned in the fastest growing segment of the core technology space: Digital Signal Processing Systems (DSP Systems). The DSP chip market was valued at about US$5 billion in the year 2000, growing at 30 per cent per annum. The market for DSP software and system design is about US$9 billion, growing at more than 50 per cent per annum.

Founders and their dreams Srini Rajam, who was the head of Texas Instruments India Ltd, and six colleagues of passionate determination decided to create a world-class technology company in India. These seven people, with 15 to 25 years of experience, came together to meet the challenge of creating 'the world's best DSP Systems Company'. TI India Ltd was one of the most innovative companies in India, as it topped the best companies operating in India that were granted US patents in the year 2000.

Venture capital Ittiam started in 2001 with a seed capital of US$5 million from Global Technology Ventures. (GTV is an investment arm of Sivam Securities, in which Bank of America has an investment stake.) After that, in the second round, the Bank of America Fund offered US$5 million for another 6.6 per cent, a price which valued this start-up at a staggering $75 million.

Products Within a year of its start, Ittiam had developed multiple products in all its target domains. This includes video imaging and audio speech products in multimedia, in addition to wireless and wireline products in communication. Ittiam also announced its wireless products, which are IEEE 802.11 based wireless LAN. Ittiam has developed solutions for both 802.11b standard which has a bandwidth of 11MBPS and orthogonal frequency division multiplexing.

Ittiam will lead the new wave of global product companies from India. The company represents the collective aspiration of the team to lead the new wave of Indian technology products, thriving in the global arena. Ittiam is singularly focused on Digital Signal Processor based systems in wireline, wireless, audio speech and video-imaging products.

Consistent with its bold vision, Ittiam is pushing at the frontiers in all the key areas: business, technology and people. In business, Ittiam has chosen to go beyond the traditional service model and has committed itself to products, both customized and off-the-shelf technology. In technology, Ittiam selected integration as its strategy algorithm, software to the actual reference board that resides in the end equipment. On the people front, Ittiam works with the fundamental belief that the company is co-owned by all who work and share the dream, irrespective of the function. The company has given shares to all its employees.

Ittiam is one of the most innovative firms operating in India, with high-quality intellectual property. DSP solution is implemented on a generic platform. Ittiam has system focus and not chip focus. The platform integrates all the interrelated domains. The company has a full-fledged marketing group and it has entered into strategic partnerships for overall solutions. In other words, Ittiam is a unique niche player with the ability to innovate. There were no technologies companies in India and Ittiam positioned itself as a technology company. The core competence of Ittiam is its capability to identify good windows of opportunity. The five aspects that distinguish Ittiam are as follows:

- experienced team,
- market focus,
- world-class orientation,
- high-level platform as the mode of integration, and
- vision to be a global leader in DSP design.

Mindtree Consulting

Mindtree is one of the fastest-growing software companies operating in India. It was selected as one of the best places to work in information technology, one of the top 100 IT employers in the USA by the third year of its establishment, according to the *Computerworld* survey in 2002. It focuses on state-of-the-art technologies and high level reusable intellectual property.

Founders A number of highly experienced people from some of the best companies got together and worked out a plan to start a new firm. The mission to deliver business enabling solutions and technologies by creating partnerships with its customers in a joyous environment for its staff. The logic of their getting together was that many of today's software services companies will not be able to be leaders in the near future, because knowledge-enabled software requires six things to remain in the leadership position: domain capability, extensive use of tools, methodology, quality, innovation and brand positioning.

Krishna Kumar was the chief executive of the Electronic Commerce Division, Wipro. Anjan Lahiri, who was working with Cambridge Technology Partners, is the second partner. N.S. Parthasarathy, general manager of Wipro's Technology Solutions is the third partner. Rostow Ravanan worked with Lucent Technologies and prior to that in KPMG. Ashok Sootha, who was the chief executive of Wipro, is the chairman of Mindtree. Kamran Ozair, who worked with Cambridge Technology Partners, was another founder. Scott Staples was also with Cambridge Technology Partners. Kalyan Banerjee, who worked as the head of Wipro Technology Solutions Division, also joined the founding team.

Vision of 2005 The company set up a very ambitious and aggressive target:

- to achieve a revenue of $231 million,
- to be among the top 10 per cent in their business, in terms of profit and return on investment (ROI),
- to be one of the top 20 globally admired companies, and
- to give a significant portion of their profit after tax (PAT) to support primary education.

Venture capitalists The first-round funding (US$9.5 million) was by the founders, Global Technology Ventures and Walden International. In August 2001, Mindtree secured the second-round funding, US$14.1 million, from Global Technology Ventures, the founders, Walden International, Capital International and the Franklin Templeton Fund.

Products and services Mindtree is essentially a services company. It operates in six thrust areas, namely, internet technologies, enterprise integration and business to business, enterprise resources planning (ERP) and supply chain management, mobile platform and technologies, application management, and setting up off-shore development centers.

The strength of Mindtree is its ability to leverage its vast knowledge base to prescribe tools and architecture which will work for specific business models and industries. The collective experience, coupled with the creation of Mindtree Labs, ensures that the solutions will have high quality and success. The focus of Mindtree, unlike that of the other software firms, has been to leverage intellectual property. Mindtree helps firms to improve their product design life cycle. Mindtree developed a set of intellectual properties to complement the product realization service offering. These technology building blocks reduce the product design cycles and may be licensed as individual reusable components. The firm has a multi-platform, multi-vendor approach to application development. It has established its own software engineering methodology, distributed rapid architecture development with quality. This methodology encompasses clear processes and measurement criteria and captures organizational learning at all the stages of product development, from concept to life cycle ownership. In three years' time Mindtree evolved into a multicultural and multinational organization. The word 'Mindtree' appears in ancient Indian literature, written in 4000BC, meaning a source of eternal intellect and wisdom for all who come into contact with it, because it springs from the mind. Recently Volvo of Sweden selected Mindtree as the global IT operations partner. Mindtree is one of the fastest-growing software firms in India.

In an interview, one of the founders indicated that companies fail, not because of the market, but because they lack experienced teams. Mindtree has one of the best teams, with strong business leadership. The focus of the company has been on intensive learning. It works with global firms and mostly on difficult projects and newer state of the art areas. The main contribution of venture capitalists has been the refinement and sharpening of the business plan.

Network Solutions

Network Solutions is a venture-funded company. It focuses on convergence solutions to network problems. It has become the preferred vendor for many firms of integrated data networks. S. Sharma, who started this, was nominated for the outstanding Entrepreneur of the year 2000 Award. The company had an income of US$3 million in 1994, increasing to US$22 million by 2002.

Founder Mr Sharma is an electronics engineer who worked with Motorola and HP for some time. Subsequent to this he initiated a number of independent projects in Asian countries, including China, India, Singapore and Thailand. While working on these projects he started a networking service firm for the multinationals operating in Bangalore. The main focus was designing networks that are cost-effective and reliable and identifying network architectures that are reliable, secure, cost-effective and platform-independent.

Venture capitalists Intel capital acquired a 15 per cent of its stake in the first round funding. This was for US$1.1 million. There was a sharp increase in its revenue after 1997. During the Internet 'bust', the management purchased the stake of Intel. Network Solutions is a now private limited company.

Products and services Network solution provision is the business of the company. This has 800 employees. CISCO, Nortel and HP Cabletron are its major clients. Network Solutions is a unique company as it is India's largest vendor-independent network infrastructure solution provider. It has become the preferred solution provider for the large banks as well as the stock exchanges in India although it was started by a single entrepreneur.

The uniqueness of the firm is that none of its customers have deserted it. The company has a prudent debt planning policy and cost management system. It has three areas of expertise and operates at eight major centers in India. It manages all aspects of the network life cycle. Recently it has started providing call centre support. It is one of five fastest growing IT companies in India according to a survey conducted by *Computer Today*. It maintains its revenue through services and retaining its client base. One of the value added services it provides is software integration. The essence of learning has been collaborative learning, which has been hierarchical. The venture support by Intel Capital helped Network Solutions by enhancing its reputation. The support provided was mostly financial in nature.

Reva

Reva is India's first electric car, designed by Reva Electric Car Company (RECC) and is the abbreviated form of Revolutionary Electric Vehicle Alternative. The vision of Reva is to establish a tradition of excellence and leadership in environment-friendly urban transport by offering the best value and highest quality electric vehicles in the world. Recently it has won an export order from the UK.

Founders Reva is the creation of the Maini group headed by Sudershan Maini. Founded in 1973, the Maini Group is today a multi-product, multi-division enterprise with business interests ranging from manufacture of high-precision products for the motor industry to electric 'in-plant' material-handling equipment, from granites to abrasives, and international trading. Sudershan Maini nurtured the idea of a small car for India for 30 years, but the idea conceptualized and took form only after Chetan Maini, his son, joined Amerigon, a USA-based company to work as a program manager on an electric vehicle project. Chetan Maini, who has a BS (Mechanical Engg) from the University of Michigan and an MS (Mechanical Engg) from Stanford University, worked for General Motors (USA) and the Amerigon Group of Incorporation (USA) before taking charge as MD of Reva Electric Car Company Private Ltd. He was the team leader of the solar car team that won the GM sun race and came third in the world solar challenge in Australia. He was also the project leader for the hybrid electric car project at Stanford University. Chetan Maini's experience with Maini precision products, his core business, which produces high-quality parts for OEMs in India and overseas, came in very handy. The group got its first taste of electric-powered vehicles at Maini Materials Movement, which manufactures high-tech equipment to transport materials and people across shop floors. The company is committed to making available facilities which offer the customer maximum comfort at a minimal cost and make Reva the vehicle of the future generation.

Origin of the idea Though the first electric vehicle was built in 1834, it was the internal combustion engine that gained popular acceptance. Gasoline-driven vehicles were faster and cheaper, with a greater range. Ready availability of petroleum products resulted in a further drawback to the growth of electric vehicles. It was only in the 1970s, when the world was hit by the oil crisis, that people realized the increasing need for alternative energy technologies for motor cars. Growing concern about environmental pollution only enhanced the interest in electric vehicles. Maini wanted to eliminate urban air pollution and he looked for new technologies that can be cost-effective. His dream was to develop the first electric car in India. The REVA project was started in 1994 and the first Reva prototype was ready in mid-1996. It was internally funded. This prototype was displayed in Bangalore in 1996–7, after extensive testing at the ARAI, Pune.

Evolution of the idea RECC is a joint venture of the Bangalore-based Maini group and Amerigon electronic vehicle technologies (A.E.V.T. Inc.) of the USA. Reva has built its reputation on leading rather than following technological change. In line with its motto to introduce technology ahead of the world to consumers in India, the company enjoys technical collaboration with world-class companies. Amerigon Electric Vehicle Technologies Inc. specializes in bringing aerospace technology to the motor industry. Curtis Instruments Inc., USA, is a manufacturer of instrumentation, controls and integrated systems for electric vehicles of all types. This company has developed the motor controller for the electric car. Tudor India Limited, a subsidiary of the largest and oldest battery company in the world (located in the USA), provided the Prestolite batteries specially manufactured for use in the Reva's high-tech power pack. Modular Power Systems USA, a division of TDI, is a world leader in charger and power supplies. The charger for Reva, which was developed by MPS, is now being made in India through a technical collaboration agreement they have with the Maini Group. The main contribution of RECC is designing, developing and manufacturing electric cars that are cost-effective and easy to manufacture.

Learning strategies Maintaining quality had always been an important issue for the Maini group. Modeled on the zero principle – zero defects, zero time delays and zero inefficiencies – the group has crafted a unique quality image for itself, both in India and abroad. Recognition of the Maini group's quality and reliability includes the ISO-9000 Certificate for three of its companies. All the components of Reva are thoroughly inspected and only after due verification are forwarded to the next stage of manufacture. Even though the first prototype of Reva was ready in mid-1996, it was introduced to the market only after extensive testing at the Automobiles Research Association of India (A.R.A.I.) in Pune for homologation.

RECC's product quality and reliability have helped it to secure several International collaborations that include General Motors, USA, and Bosch, Germany.

R&D strategy The Maini group has always viewed technology and innovation as the main drivers of growth and profitability. The group has always focused on innovation, technology, quality and reliability. It has two inhouse R&D centers, recognized by D.S.I.R. (Dept of Scientific and Industrial Research, Govt of India). Reva has a 25-strong R&D team which is constantly striving to improve product quality. It is working to come out with a bus by the end of 2003. The company is also working on an enviable project of drive system for General Motors. Keeping in step with international standards, Reva spends almost 8 per cent of its turnover on R&D. The R&D efforts have resulted in innovative technologies that are patented. Apart from its design, Reva deploys three key patent-protected technologies in its electric car: running chassis, energy management system, and climate control system.

Market dynamics Most of the capital equipment, except for a few sophisticated machines, is indigenously available. The battery could be charged using a 220-volt, 15-amp power source; the payload is 227kg. Reva was developed as a completely indigenous car for India. Unlike the conventional internal combustion engine car which has more than 7000 components, Reva has only 1000 components, and more than 95 per cent of these components are indigenously manufactured. Examples where RECC used its manufacturing philosophy innovatively include the use of color-impregnated panels to eliminate any painting at the assembly stage. This construction method reduced capital costs by 40 per cent. Opting for a thermo-formed (rather than injection-molded) instrument panel, and dispensing with curved glass and winding windows, the makers selected conventional lead-acid batteries rather than new-generation lithium types. Reva has entered into a deal with 'Going Green', a UK company, for 250 000 Reva cars over the next 10 years.

Institutional support Reva received commendable support from the Department of Information Technology, IISc., Bangalore. It also receives support from Maini Info Solutions, a subsidiary of the Maini Group. On the financial front, Reva received financial support from the Technology Development Board (India). According to the company, government support for the electric vehicle industry is not adequate. It is appropriate that this venture receive the support of the government, since the technological performance of the electric vehicle largely meets the specifications. The Technology Development Board gave RECC a new venture loan of RS185 million for development and manufacturing.

Organizational strategy The marketing strategy is aimed at developing a whole new concept in city mobility: non-polluting, noiseless, affordable personal transport for all ages. The company's target is to sell 1500–2000 cars in 2003–4. According to Maini, electric cars are appropriate in city environments thanks to increased mobility, zero pollution, less parking space and quiet operation, and it is particularly tailor-made for countries like India because of low running and maintenance costs. The feedback shows that, for most buyers, the Reva is their second car, which they prefer to use in-city, while their regular vehicle is used for long-distance trips. The company is also working on a platform for larger electric cars. It is planning to divest 25 per cent equity at premium.

The deluxe version and AC version were launched in 2002. A version with a longer range than the present is in the pipeline. The 75-strong R&D team at RECC is also working on a car with a heating option and another one with cooled seats. There are also plans to expand the Reva platform by launching another vehicle by 2003. In the five years since its inception, the Reva project has cost US$20 million, with an additional US$5 million to put the car into production. ICICI has invested about US$3 million in Reva and the Maini Group has an investment of US$20 million. Reva has got an export order from Europe for 100000 cars for the period 2003 to 2008.

Features of the REVA car are as follows:

- running cost of 40 paisa per km,
- priced at Res 254000,
- zero pollution,
- seat two adults and two children,
- easy driving as it has no clutch or gears,
- on a single charge, the Reva can be driven for 80km,
- two-door hatchback,
- battery lifespan of 40000km, which should last for three or four years in city driving conditions.

Learning from the case study This study on Reva gave an understanding as to how a company could leverage technology to develop world class products indigenously. This innovative creation from the Maini group was helped tremendously by Chetan Maini's previous experience in electric vehicular technology. This is one of the biggest funded projects that is supported by TDB, which is a quasi-venture fund operated by the government of India.

Shantha Biotechnics

Shantha Biotechnics is engaged in the development, production and marketing of biotechnology-based human healthcare products. It has developed a Hepatitis B vaccine, Shanvac. Shantha Biotechnics was the first Indian company in to use recombinant DNA to create a pharmaceutical product. It is the first indigenously produced vaccine for Hepatitis B. With an estimated 42 million Hepatitis B vaccine carriers – a whopping 4 per cent of the country's population – India is the second-largest pool of carriers in the world.

Founders The man behind 'Shanvac' is Varaprasad Reddy, an electronics engineer by training. He worked in the Defense Electronics Research Lab, then started a battery-making unit for supplying high-power batteries to the Indian Airforce. He had the urge to do something for India and also the urge to be an entrepreneur. He wanted to start a new industry that is more challenging. Both innovation and entrepreneurship were his dreams. When Reddy went to the USA, people suggested that he should focus on biotechnology as there are many new opportunities emerging. When he attended a workshop in Europe someone mentioned the need for vaccines in developing countries. This immediately became a trigger for action, and he worked relentlessly. His dream was to introduce affordable products that can have a significant impact.

Venture capitalists Reddy was looking for a venture capitalist. The foreign minister of Oman, H.E.Yusuf Bin Alwai, visited Shantha Biotechnics when he came to Hyderabad. He liked the project and invested US$1.3 million as an angel investor. Then the project took off. Meanwhile, the Technology Development Board gave a loan of Re85 million (US$1.7 million) as the first stage, and SBI Mutual Funds invested US$11 million and acquired a 6.9 per cent stake in Shantha Biotechnics. Subsequently, TDB again made a loan of Re180 million. In 2002, Shantha Biotechnics achieved a sales turnover of Re300 million (US$6 million).

Product and services Shantha has a state-of-the-art facility equipped with sophisticated instrumentation for industrial R&D in modern biotechnology. The company is the largest private sector biotechnology company in India. Shantha developed India's first genetically engineered r-DNA Hepatitis-B vaccine after five years of intense research. It developed India's first genetically engineered Interferon alpha 2b Shanferon. The vision of Shantha Biotechnics is to achieve breakthroughs in modern biologicals, leading to development of products and services that address critical healthcare needs at affordable cost. The company has 376 employees of which 75 are R&D personnel. It commercialized streptokinase in the last quarter of 2002. In the next two years it will commercialize recombinant Erthyropoeitin, insulin G-CSF and GM-CSF.

Shantha Biotechnics has a joint venture with East West Labs, USA, for the development of novel therapeutic monoclonal antibodies for the treatment of different types of cancer. The targets are non-small cell lung cancer, breast cancer, pancreatic cancer, colon cancer and melanoma, for which it has patents. It is a firm that is intensifying its drug discovery efforts.

Strategies and learning Shantha Biotechnics moved quickly in the drug discovery cycle, through intensive learning. It has four technological alliances that facilitated learning:

- Shantha Marine Biotech has a joint venture with ABL Technologies to focus on marine biotechnology products.
- Shantha Biotech has tied up with Pfizer for marketing Shantha's products locally in India and, in future, in the global markets.
- It has a research alliance with the International Vaccine Institute, Korea, for the technologies for a typhoid vaccine.
- Shantha Biotechnics has formed a joint venture for producing human insulin costing US$5 million.

Shantha Biotechnics is planning to go for an IPO in the near future. Before that it is building the product pipeline. For manufacturing, Shantha Biotechnics has entered into a joint venture agreement with Biocon, located at Bangalore. The venture capitalist helped in getting a large investment from a bank for the building of manufacturing facilities, against his personal guarantee.

Kshema Technologies

Kshema Technologies was founded in 1997. It is one of India's first venture capital-funded software solutions companies. With an annual growth rate of more than 125 per cent since inception, Kshema is the country's fastest growing software company and has clients predominantly from 1000 global companies.

Founders Kshema was promoted by A.R. Koppar, A.Mutalik and L.B. Joseph who came together to realize a dream of creating an employee-owned organization. The first two are engineers. They worked earlier at Wipro at senior levels. The dream was to create a firm that is innovative and ethical.

Venture Support The three investors who invested in Kshema are Global Technology Ventures, IL & FS Venture Capital Corporation Ltd and Citibank. Global Technology Ventures have bought a 50.88 per cent stake, IL & FS Venture Capital Corporation Ltd 12.69 per cent, Citibank holds 4.61 per cent. The software revenue in 2002 was Re560 million. Profit after taxation was Re122 million. In spite of the poor markets, revenues have not shown any substantial decline.

Product and services Kshema's Software Services is a firm that has 45 clients from the Global 1000 (*Business Week*) firms. Its mature software development is backed by years of experience in delivering software solutions in a global environment. The services are based on object technologies, web technology, wireless solutions and automation. Automotive embedded software, in-vehicle multi-media systems, embedded technologies for hand-held devices and so on are some of the major technologies of Kshema. The company has been at the forefront of handheld device evolution and has been involved in some of the world's first technology prototypes in this area. The pioneering work Kshema has done includes

- integration of the Bluetooth communications module for personal digital assistant (PDA),
- design and development of a new generation of PDA and phones,
- bluetooth stack for hand-held devices,
- voice recognition integration for new generation devices, and
- word processor and spreadsheet applications for a PDA platform.

Kshema has recently started providing bioinformatics services. Stimulation of metabolic pathways using databases, datawarehouse application for genetic sequencing and so on are some of the applications. Kshema has recently developed an image-enhancing and spot identification system to map coordinates of the protein shots for robot excision.

Kshema has strong motivation systems for inducing learning. It operates on a customer-centric 'virtual extension' business model that ensures value at every stage in its software development cycle. The three venture firms have been able to increase the value creation in three ways, namely (a) by continuous monitoring of business plans, (b) by helping to get business contacts from different countries, and (c) providing funding quickly. In a shortest possible time, the venture came into operation as one of the venture capitalists provided the necessary infrastructure.

Impulsesoft

Impulsesoft is an innovative design firm that has developed 802.11 products for Japanese firms. Founded in 1998, Impulsesoft is a venture-assisted firm that was supported by the most renowned angel investor in India: N.S. Raghavan. He was one of the co-founders of Infosys. Impulsesoft is headquartered in India, with branches all over the world. It is focused on delivering short-range wireless solutions including Bluetooth and 802.11.

The management team The firm was started by three well-known experts, Chandrasekharan, Srikrishna and S. Bhaskar. They came together to provide innovative products to various users. The management team is supported by an advisory team of well-known venture capitalist experts, who bring in an enormous amount of experience.

Angel investor Mr Raghavan, who co-founded the software major Infosys Technologies, invested in this company as an angel investor. Impulsesoft has a unique and innovative product portfolio and Mr Raghavan brings with him an enormous reputation and excellent insights into the intricacies of software business. All the firms in which he has invested are considered star firms by the analysts. Two globally reputed venture capitalists, Motti Beck and Gunnar Hurtig, are on the advisory board.

Products The wireless product portfolio of Impulsesoft consists of Bluetooth Protocol Software, Reference Design and development kits. Impulsesoft has supplied a very innovative product solution to Matsushita, Japan.

Partnerships Impulsesoft has formed strategic partnerships with various leading Bluetooth vendors and technology companies such as Broadcom, Cirrus logic, Infineon, Microsoft, National Semiconductor and Silicon Wave. These global partnerships with reputed firms have been a source of strength for Impulsesoft.

Support of the angel investor The angel investor has provided guidance so that the business plan is regularly carried through. The support of a highly reputed and experienced angel investor along with a very dedicated team ensures that this firm meets its business obligations. The angel investor helps in balancing technology and business risks.

Mitoken

Mitoken was established in May 2000. It offers a new class of web-enabled enterprise applications for managing IT and software enterprises. The name, Mitoken, is derived from Mitos, Greek for strands, and Ken, English for knowledge. Mitoken has strategic sponsorship from the Global Software Group of Motorola and venture funding from India's largest venture fund, ICICI Ventures.

Management team The team consists of a group of four people, three of whom came from Motorola. The team has 50 years of consulting experience in software. The team, Srinivas Pannala, Shishir Pathak, Seshadri Iyer and Bhoopalan Padua, wanted to create a world-class product company and so came together.

Vision The vision was to create a world-class product company with a deep understanding of clients and sharp execution skills. Accordingly, it provides a creative climate. The vision focuses on attracting and engaging the best people. To realize the vision the company uses engineering excellence, reliability and sound technology infrastructure as enablers.

Products Mitoken developed four products that can help software business to enhance its productivity:

- Project portfolio management: for managing software projects.
- Product budgeting and profitability: for assessing profit margin on a regular basis.
- Opportunity management: improving sales pipeline visibility.
- Contact management: initiating, deepening and managing relationships.

The company has recently developed a workflow engine, the Mitoken Business Workflow Engine, which is the best in its class.

Origin of the firm While working with Motorola three of the founders developed a software product. This was used by most Motorola divisions across the globe. As the product was good, the team thought of incubating it. The head of Motorola India fully supported the move. He was convinced that the team had the necessary commitment and drive to become a new firm. As the developed software was the intellectual property (IP) of Motorola, the Technology Transfer Review Board of Motorola (global headquarters) had to clear the transfer to a new firm. They asked for the views of all the internal software divisions across the world and the transfer was agreed upon. This was the origin of Mitoken. The initial incubation expenses were borne by Motorola and the firm was originally located at Motorola. The three critical success factors have been executive sponsorship by Motorola, the drive and commitment of the team and the rich experience of the founders.

Lessons Incubation can be a good way of creating new ventures but prior experience is a crucial determinant of commercial success. Incubation support by a global firm (Motorola) gave the venture the necessary initial support for its growth.

INSIGHTS FROM THE CASE STUDIES

The venture capital industry has started creating innovative firms in India. Over the last five years many new entrepreneurial firms have ventured into new product development and contract research for global firms. Previously Indian firms had been weak in new product development. Firms like Avesthagen, Strand Genomics and Bharat Biotechnics have achieved high revenue levels through contract research as well. Firms like Tejas Networks, Reva and Ittiam have become product developers for the global market. Mindtree and Kshema have grown rapidly by focusing on new high-technology business segments. Venture capital-assisted firms are still in their infancy. Management buy-outs and external corporate venturing have begun to emerge, indicating that off-shore funds are beginning to consider India as presenting a potential opportunity. This will reduce the capital gap for entrepreneurial firms. A summary of the insights from the cases is given in Table 8.6. Major observations are given below:

1. Venture capital is becoming a major mechanism for stimulating innovation and entrepreneurial growth. In India, this is catalyzed by the rapid growth in information technology. There is a strong need to enhance availability of venture capital in developing countries as most of these at risk-averse, but awareness about the role of venture capital has been very limited. There have to be systematic initiatives for simu-

Table 8.6 Insights from the cases

Firm	Insights
1. Tejas Networks	Reputation enhances venture funding opportunities Creating local and global customers enhances sustainability
2. Strand Genomics	Venture support helped in developing new markets Developing long-term alliances minimizes the business risk
3. Avesthagen	Contract research provides for a steady stream of revenues and it can help in commercialization of innovations
4. Ittiam	Bringing together a highly experienced team and concentrating on a niche segment can crate high value ventures
5. Mindtree	Clear business plan supported by strong business leadership attracts venture support
6. Network Solutions	Corporate venture capitalists help in building reputation
7. Reva	Venture support and prudent management can help in managing business risk
8. Shantha Biotechnics	Venture capital helps in mobilizing larger chunks of capital
9. Kshema Technologies	Business plans were rigorously assessed and monitored and this led to rapid growth in the second phase
10. Impulsesoft	Experienced and reputed angel investor can help a firm in the early stage to ensure success
11. Mitoken	In incubator, support of executive sponsor is essential

lating entrepreneurship through use of venture funds. The distortions in the capital market due to overregulation and multiple controls are also a problem that is hindering the growth of VCs.

2. Expertise needed for managing new ventures and managing venture funds has yet to evolve in India. Most of the off-shore funds have a strong experiential base that is absent in local institutions. Off-shore funds have been able to provide support and business contacts. From personal interviews it is evident that off-shore funds are able to add more value to the venture-assisted firms through the provision of help in preparing reliable and precise business plans. Entrepreneurs generally focus on technical aspects and not on business success. Venture capitalists bring a balance between business and technology so that innovation becomes a commercial success.
3. Most of the new ventures have benefited from venture capital, especially those supported by the off-shore funds. Four aspects of support

provided by VCF that adds value are: monitoring of the business plans, support for getting business contacts from other countries, bringing an external perspective to the business plan, and enhancing the reputation of the firms.

4. Venture capital growth and industrial clustering have a strong positive correlation. Foreign direct investment, the starting of R&D centres, availability of venture capital and growth of entrepreneurial firms are becoming concentrated into five clusters. The cost of monitoring and the cost of skill acquisition are lower in clusters, especially for innovation. Entry costs are also lower in clusters. Creating entrepreneurship and stimulating innovation in clusters have to become a major concern of public policy makers. This is essential because only when the cultural context is conducive to risk management will venture capital take-off. Clusters support innovation and facilitates risk bearing. VCs prefer clusters because the information costs are lower. Policies for promoting dispersion of industries are becoming redundant following economic liberalization.
5. An analysis of venture-assisted firms clearly shows that the factors contributing to the success of innovative firms are essentially threefold, namely strong experiential base, vision and the urge to achieve something, and a realistic business plan.
6. Bank-operated venture capital funds are relatively risk-averse and they have a weak experiential base. Local funds are focusing on software services and retail business but not innovative products. The real growth of venture capital in India started after the entry of off-shore venture funds. India has become a preferred destination for venture funds in Asia.
7. The presence of an excellent academic research institution is a prerequisite for the success of venture firms in a location as it can provide high-quality manpower. In the case of Mumbai, Madras, Hyderabad, Bangalore and New Delhi, the presence of research institutions has facilitated the growth of venture-supported firms.
8. One of the untraded externalities that stimulates venture growth is idea entrepreneurship. An idea moves faster and evolves quickly in clusters. Venture capital growth has occurred in clusters in India as in the USA, Israel, the UK and Taiwan.
9. In developing countries, venture funds are not fully evolved and it may be necessary to start public venture funds. Public venture funds can act as seeds of entrepreneurship. Special attention may be essential for this so that commercial and technical perspectives are integrated. In developing countries, public policy should support and evolve institutional systems for stimulating public venture funds. The government-

supported quasi-venture fund, the Technology Development Board, has been effective in stimulating innovations in India. Good corporate governance of venture funds is one of the critical success factors that has helped the Technology Development Board to select and support innovations.

To sum up, developing countries have to harmonize the capital market requirements and venture capital needs so that they can stimulate entrepreneurial firms that focus on high-tech innovations. Though most venture funds state that high technology is their priority, only firms started by experienced people find support by VCFs. The capability to assess venture projects continues to be a weak area in the case of developing countries such as India because of the lack of prior experience.

NOTES

1. According to the report of the Committee on Venture Capital (Chairman: K.B. Chandrasekhar), this multiple set of guidelines has created inconsistencies and detracts from the overall objectives of development of VC industry in India as all three sets of regulations prescribe different investment criteria for VC funds. The report can be found at http://www.sebi.gov.in/report/venture/exesumm.html.
2. According to the AVCJ classification, the category 'computer-related' includes both hardware and software and 'information technology' includes Internet related activities.
3. Carrier class is the optic fibre cable approved for long distance cable application.

BIBLIOGRAPHY

Asian Venture Capital Journal (various issues), *The Guide to Venture Capital in Asia*, Hong Kong: Asian Venture Capital Journal.

Bowonder, B. (2001), 'Globalization of R&D', *Interdisciplinary Science Review*, **26**(3), 191–203.

Chitale, V.P. (1983), *Risk Capital for Industry*, New Delhi: Allied.

Dossani, R. and M. Kenney (2002), 'Creating an Environment for Venture Capital in India', *World Development*, **30**(2), 227–53.

Gompers, P. and J. Lerner (2002), *The Venture Capital Cycle*, Cambridge, MA: MIT Press.

Government of India (2002), *Economic Survey, 2001–02*, New Delhi: Ministry of Finance, http://www.indiabudget.nic.in/es2001-02/welcome.htm.

IVCA (2000), *VC Industry in India*, Bombay: IVCA.

Learner, J. (1999), 'The Government as Venture Capitalist: The Long-Run Impact of the SBIR Program', *The Journal of Business*, **72**(3), 285–318.

Learner, J. (2002), 'Venture Capital', in B. Steil, D.G. Victor and R.R. Nelson (eds), *Technological Innovation and Economic Performance*, Princeton: Princeton University Press, pp.327–46.

Mani, S. (1994), 'Financing Technology-based small firms and venture capital funds: the Indian Experience', Centro Studi Luca d'Agliano – Queen Elizabeth House Development Studies Working Papers, no.78.
Mani, S. (1997), 'Financial Barriers to Domestic Technology Development and Venture Capital', *Science, Technology and Development*, **15**(2 and 3), 227–40.
McNally, K. (1997), *Corporate Venture Capital*, London: Routledge.
National Science Board (2002), *Science and Engineering Indicators*, Washington, DC: National Science Board.
Premus, P. (1985), *Venture Capital and Innovation*, Washington, DC: USGPO.
Ramesh, P. and A. Gupta (1995), *Venture Capital and the Indian Financial Situation*, Bombay: Oxford University Press.
Smith, R.L. and J.K. Smith (2002), *Entrepreneurial Finance*, New York: Wiley.
Verma, J.C. (1997), *Venture Capital Financing in India*, New Delhi: Response.
Verma, Y. (2002), *Dataquest*, **20**(6), 42–8.

9. The emergence of the Singapore venture capital industry: investment characteristics and value-added activities

Clement Wang

INTRODUCTION

In this chapter we examine the Singapore venture capital (VC) industry in three sections. The first section starts by looking at the historical evolution in four phases. We describe the various policies and programs set up by the Singapore government in supporting the VC industry as well as tracing the role it has played over the years. The second section focuses on the current structure of the VC industry in Singapore. Here, various VC industry trends and organizational characteristics are profiled. Next, from a case study of ten VC firms in Singapore, we group the value-added activities by these firms into four categories: technological, marketing, management and financial support, and summarize their similarities and differences. Finally, the chapter concludes with a discussion on VC firms' governance structures which characterize their value-added activities to their investee companies. We also draw some managerial implications for VC firms and policy makers.

HISTORICAL EVOLUTION OF VENTURE CAPITAL IN SINGAPORE

Singapore is one of the most active players in venture capital among Asian countries and has established itself as a regional financial hub. The country's VC industry began in the early 1980s and developed rapidly in the 1990s. At the end of the year 2000, the cumulative VC under management in Singapore stood at US$9.2 billion, ranking it as the third largest VC investment pool in Asia, next only to Hong Kong and Japan (AVCJ, 2002). The evolution of venture capital in Singapore can be divided into four phases, as shown in Table 9.1.

Table 9.1 The evolution of venture capital in Singapore

Phase	Years	Milestones
Inception phase	1980–85	creation of the Singapore Science Park (1980)
		first VC fund raised (1983)
Early development phase	1986–90	creation of the EDB VC Fund (1986)
		birth of SESDAQ (1987)
		total VC pool in Singapore reached S$1 billion
Expansion phase	1990–96	foundation of EDBI (1991)
		formation of the Singapore Venture Capital Association (1992)
		establishment of innovation programs (1996)
New development phase	1996–2002	launch of Technopreneurship 21 (T21) (1999)
		launch of the Startup Enterprise Development Scheme (1999)
		launch of the Technopreneurship Investment Fund (TIF) (1999)
		launch of Venture Investment Support for Start-ups (VISS)
		establishment of the OCBC TechFinanching Centre
		forming of the Entrepreneurship and Internationalisation Sub-Committee (2001)

Phase 1 Inception Phase (1980–85)

The inception of the VC industry in Singapore took place in the early 1980s with the entry of the SEAVI (South East Asia Venture Investment) program and the creation of the Singapore Science Park (SSP), the island's own 'Silicon Valley' or 'Technology Corridor'. SEAVI was established in 1983 by one of the world's major VC groups, TA Association (USA) and Orange Nassau, a Dutch investment management firm (Chin, 1988; Chia and Wong, 1989). In the same year, SEAVI raised its first VC fund, also the first VC fund in Singapore, which amounted to S$14 million. It focused on investments in small and medium-sized enterprises (SMEs) with high growth potential. In the following two years, SEAVI raised two other funds, with the total VC under management by SEAVI reaching S$97 million, the total VC pool in Singapore at that time.

As another identification indication of the inception of the VC industry in Singapore, the Singapore Science Park was set up under the government's initiative in 1980. Noting the success of Silicon Valley and other high-tech parks in the USA, the UK, Canada, Japan and Korea, the Singapore government – specifically, the Singapore Science Council and the

Economic Development Board (EDB) – decided to set up a science park in Singapore to provide new infrastructures to promote R&D activities and entrepreneurship. The establishment of the Singapore Science Park ensured the easy tapping of a vast pool of investable funds and venture capital for entrepreneurs and young companies.

Phase 2 Early Development Phase (1986–90)

After the initial phase, the Singapore government began to take steps to nurture the venture capital industry from 1985 onwards. Through active government support and promotion, the VC industry in Singapore experienced an early growth from 1986 to 1990. At the end of this period, the total VC pool in Singapore amounted to more than S$1 billion.

The role of EDB

Since 1961, in addition to attracting foreign direct investment and wooing MNCs to come to Singapore, the Singapore Economic Development Board (EDB) been acting as a catalyst and facilitator to nurture entrepreneurship in the country, and has played an important role in the development of the VC industry in Singapore. Generally, EDB's role in VC can be described in two aspects: direct investments in VC and setting up strategic programs/schemes to promote VC development in Singapore.

Direct investment in VC At the end of 1985, EDB created its first S$100 million VC Fund. Under the EDB VC fund scheme, when the project is successful, the investees can purchase EDB's equity stake within a specified period at a fair market price (Chia and Wong, 1989). In the period 1986 to 1990, the EDB invested in seven funds in Singapore and overseas, and in more than ten companies (with a total portfolio of more than S$33 million) in the high-tech arena (Buchanan, 1989).

Setting up strategic programs/schemes Besides its role in direct VC investment, EDB has also set up various strategic programs or schemes to promote VC development in Singapore. During the period of early development from 1986 to 1990, it established the Venture Capital Club and the Strategic Business Unit to help local VC firms through important programs or schemes such as the Capital Assistance Scheme.

The EDB undertook the role of a broker through the creation of the Venture Capital Club (VCC) in 1986. In essence, the VCC endeavored to bring together investees and their potential financial sponsors via such channels as monthly meetings or forums. In mid-1988, the EDB formed the Strategic Business Unit (SBU) to identify overseas investment opportunities.

The formation of SBU was timely as it enabled Singapore VC firms to gain access to overseas investment opportunities. EDB also sponsored a special loan scheme, the Capital Assistance Scheme, to promote investment in high-tech industries. This special credit scheme started after the 1984/5 recession, when EDB managed to obtain a S$200 million credit loan from the Ministry of Finance (Chia and Wong, 1989). The amount increased to S$300 million towards the end of this phase.

In addition, the birth of the SESDAQ (Stock Exchange of Singapore Dealing and Automated Quotation system) market in 1987 was another important milestone of VC development in Singapore. As a market to complement the SES (Stock Exchange of Singapore), SESDAQ provided an avenue for both the venture capitalists and the entrepreneurs to seek capital appreciation or profits from their investments, by relaxing the entry criteria and listing requirements (see Table 9.2). Currently, the SESDAQ has 113 companies, with a market cap of S$3.5 billion, not including the 51 that have since transferred to the mainboard.[1]

Table 9.2 Some major admission criteria of SESDAQ

Time length	The company must have been trading for at least three years
Pre-tax profits	Business is expected to be viable and profitable, with good growth prospects
Track record	A company with no track record has to demonstrate that it requires funds to finance a project or develop a product, which must have been fully researched and costed
Shareholding spread	At least 500 000 shares or 15 per cent of issued shares (whichever is greater) in the hands of at least 500 shareholders
Continuing listing obligations	Yes, but waiver obtainable if company is also listed on another recognized foreign stock exchange

Source: SGX (2002).

Besides EDB support and SESDAQ, the Singapore government also introduced several other packages of tax and non-financial incentives for VC investment, which included the following (compiled from: Chia and Wong, 1989; EDB, various years):

1. Investment allowance – companies that planned to invest in R&D and entrepreneurial activities could apply for an investment allowance of up to 50 per cent of the investment.
2. Tax-free capital gain – the Singapore government regarded gains from

the disposal of long-term capital investment by VC businesses as tax-free gains.

3. Venture capital incentive – This incentive enabled investors to write off against their other income, up to the full amount (100 per cent) of capital loss arising from the disposal of shares in an approved project. The philosophy of this incentive was to encourage local entrepreneurs to invest in higher-risk technologies and ventures.

Phase 3 Expansion Phase (1990–96)

After the initial development of the VC industry in the 1980s, Singapore experienced the booming development of venture investment in the 1990s, especially before the Asian financial crisis in 1997. From 1990 to 1996, Singapore experienced an average increase of about 30 per cent in VC pools. During this period, government and non-government incentives propelled VC development to new heights.

EDB Investments (EDBI)

The Singapore government continued to promote the VC industry through the functions of EDB. Among EDB's programs during this period, the most important incentive to VC firms was provided through EDB Investments Pte Ltd (EDBI). Wholly owned by EDB, EDBI was set up in 1991 as an independent investment equity aimed at accelerating the growth of enterprises and industry clusters, and promoting emerging industries and innovative technologies.

Currently, the EDBI directly manages ten funds: the Cluster Development Fund (CDF), EDB Ventures (EDBV), EDB Ventures 2 (EDBV 2), PLE Investments (PLEI), the Mobile Commerce Venture Fund (MCVF), the BioMedical Sciences Investment Fund (BMSIF), the Pharmbio Growth Fund (PGF), Life Sciences Investments (LSI), Singapore Bio-Innovations (SBI) and the Technopreneurship Investment Fund (TIF). EDBI also co-manages the Smart Lab 'Incubator Fund' with a corporate VC firm.

Up to the end of 2001, EDBI managed funds in excess of S$6 billion (US$3.3 billion) and invested in more than 280 projects covering a diverse range of businesses in Singapore and internationally (EDB, 2002). Almost 70 per cent of EDBI's direct investments were in seed or early-phase companies in growth sectors such as communications (including wireless applications), software/industrial services, electronics and biomedical sciences.

Singapore Venture Capital Association

To assist the development of venture investment, the Singapore Venture Capital Association (SVCA) was formed in 1992 as the industry's platform to foster growth, raise awareness, promote standards and act as a source for networking. Currently the SVCA has 40 full members, with 45 associated corporate members and individuals (SVCA, 2002). The SVCA does not focus solely on Singapore but also looks into regional development and global VC trends. Over the years, as the industry platform, the SVCA has performed notable roles and functions. These include (compiled from SVCA, 2002):

- being a regular supporter of the government's 'TechVenture' series of annual conferences for VC firms and technology start-ups (since 1997);
- setting up ties with the Singapore American Business Association to promote business relationships between the United States and Singapore (1997);
- organizing meetings with the Stock Exchange of Singapore for the VC industry's views on relevant issues (1997);
- setting up a Corporate Finance Committee in 1998 to make recommendations to the Monetary Authority of Singapore regarding the liberalization of Singapore's capital market;
- co-leading, with the Hong Kong VC Association, in forming the Asia Pacific Venture Capital Alliance (APVCA), a regional platform to promote further development and growth of the VC/private equity industry, in November 2001.

Innovation programs

In 1996, EDB first established the innovation program to spearhead the promotion of innovation among companies and thus stimulate venture investment. The objective of the program was to promote a pervasive sense of change and growth among the manufacturing and related services sectors in order to develop a wide base of indigenous creative capabilities in Singapore. One key initiative was the setting up of the Innovation Development Scheme (IDS) in 1996 to encourage companies from the manufacturing and services clusters to undertake innovation projects in Singapore, as the anchoring of product and process charters in Singapore is an important measure that encourages sustainable innovation to persist. Since 1999, EDB has attracted companies to establish ten product/process charters in Singapore (EDB, 2001). Through seminars, emphasis is also placed on creating opportunities for companies to undertake innovative projects with leading practitioners in the area, who will indirectly encourage them to adopt well-proven innovative systems and methodologies within their companies. To further enhance Singapore's innovation environment, EDB, together with the Productivity and Standards Board (PSB),

Ministry of Finance (MOF) and the Public Service Division (PSD), coorganized national-level Innovation Awards, the highest level of recognition for innovation. These awards are intended to reward deserving individuals and organizations for their exemplary contributions in the area of innovation. Internationally recognized personalities in the field of innovation are invited to sit on the panel of international judges to strengthen the message of the importance of innovation.

Phase 4 New Development Phase (1996–2002)

The Asian financial crisis and the ensuing economic recession have severely affected the development of the VC industry in Singapore. Although the overall VC pool was still increasing during this period, both the new funds raised and the amount invested decreased. To address the situation, the Singapore government launched several programs to promote VC investment and entrepreneurship.

Technopreneurship 21

Launched in April 1999, Technopreneurship 21 (T21) was an initiative involving high-level government and private-sector efforts to prepare and lay the foundation for the successful development of a technopreneurial sector in Singapore. Because investments in high-tech start-ups are generally risky, the government encouraged investment into such high-tech start-ups by sharing the risk with investors by introducing the Technopreneur Investment Incentive (TII). Qualified technopreneur start-ups would be given TII status by EDB. In the year 2001, 40 TII applications were approved (EDB, 2001). The start-ups could then issue certificates to its investors of up to a maximum investment of S$3million. Investors with valid certificates were then entitled to deduct their loss amounts against their taxable income.

Startup Enterprise Development Scheme

Also launched in 1999, the Startup Enterprise Development Scheme (SEEDS), a S$50 million fund administered by the EDB, aimed to provide equity financing for start-ups in the seed phase. Under the scheme, EDB would match, dollar for dollar, every private dollar raised from third party investors by the seed/start-up company, up to a maximum sum of S$300000. Both EDB and the third party investors would take equity stakes in the company in terms of their investment percentages. Successful SEEDS applicants spanned a wide spectrum of industries: IT, biotechnology, electronics and e-commerce; media and communications; and nanotechnology.

Technopreneurship Investment Fund

Technopreneurship Investment Fund (TIF) Ventures is also a corporatized entity wholly owned by EDB to manage *fund of funds* and associated activities. TIF emphasizes the harnessing of new technologies and focuses geographically on the USA (East and West Coast), Europe (Central, Nordic and Eastern), Israel, Korea/Japan, India, China, Singapore and Australia/New Zealand. Currently, TIF is a US$1 billion VC fund of funds that seeks to help Singapore-based institutions to participate in the local VC industry. TIF is made up of three tranches, including a US$250 million early-stage fund, a US$500 million broad base fund, and a US$250 million strategic fund (EDB, 2002). The objectives of TIF are as follows:

- to obtain superior financial returns from investments in globally diversified top-tier VC funds;
- to coinvest with TIF fund partners in private companies that offer substantial returns (up to 20 per cent of total fund size);
- to invest in indigenous VC funds that have shown excellent potential (up to 10 per cent of total fund size).

Venture Investment Support for Start-ups

Venture Investment Support for Start-ups (VISS) is a SGD$50m coinvestment program that coinvests directly in early-stage promising and strategic companies that are based in or linked to Singapore. It aims to play a catalytic role in drawing investments into earlier stage start-ups by lowering the mental barrier through lower risk exposure. TIF Ventures Pte Ltd would invest in a company with a minimum leverage factor of $1 of VISS investment for every $2 of private investment. TIF Ventures stipulates that its investment amount will not exceed S$500000 and that it will not become the largest single shareholder in the company.

OCBC TechFinancing Centre

The first bank to offer venture banking in Singapore, OCBC established the OCBC TechFinancing Centre to offer credit facilities to technology start-ups which have already raised capital from venture capitalists or qualified business angels. The credit terms cover items such as working capital, equipment and accounts receivable financing. In addition, collateral is not always required for these credit facilities.

Entrepreneurship and Internationalisation Sub-Committee

To review fundamentally its development strategy and to formulate a blueprint to restructure the economy, the Singapore government set up an economy recession committee under the Ministry of Technology in 2001.

The committee set up seven subcommittees, including one named the Entrepreneurship and Internationalisation Sub-Committee (EISC), focusing especially on venture investment and entrepreneurship in Singapore. The terms of reference for the EISC are to recommend ways to strengthen the spirit of entrepreneurship and innovation in Singapore, and to foster the growth and internationalization of Singapore-based companies. The framework of recommendations by the EISC includes six Cs: culture, capabilities, conditions, connections, capital and catalysts (compiled from ERC, 2002):

Culture The EISC recommends creating opportunities for young people to develop strong entrepreneurial instincts and understanding. The suggested approach is to provide all students at different levels, as well as working executives, the opportunity to learn about enterprise and gain early exposure to business concepts.

Capabilities The EISC recognizes that enterprise capabilities can be developed at both the individual and the industry level. It thus recommends that, at the individual level, Singapore should undertake a serious effort to attract global entrepreneurial talent, as well as to tap local talent. It also recommends that, at the industry level, there is scope to foster greater collaboration between enterprises to leverage each other's strengths to venture abroad.

Conditions The EISC recommends that the government should take enterprise-friendly approaches to economic management, industry regulation and procurement. The EISC also suggests some principles and approaches for the government's involvement in the private sector to help develop conditions for a more entrepreneurial economy.

Connections The EISC feels that Singapore can improve much in terms of soft infrastructure to connect people and markets, and hence recommends outward-oriented policies to forge strong linkages between foreign markets and its Singapore home base.

Capital The EISC recommends that mechanisms be put in place to encourage more private sector financing of enterprises at start-up, growth and internationalization phases, and for the government to implement broad-based measures to free capital tied up in the Central Provident Fund (CPF) and in Housing Development Board (HDB) properties.

Catalysts The government has been using incentives to attract multinational companies (MNCs) to invest in Singapore and drive local economic growth. To develop another key driver for Singapore's growth, it is considered necessary to encourage enterprise and wealth creation. The EISC therefore recommends a package of incentives and changes to tax regulations to channel more capital towards enterprise.

CURRENT STRUCTURE OF THE VC INDUSTRY IN SINGAPORE

Since the birth of the VC fund in Singapore in the early 1980s, the VC industry has grown significantly, both in the number of VC firms and in the amount of funds, in the 1990s. Although the Asian financial crisis and the succeeding economic recession have slowed the pace of VC development, the Singapore government managed to promote VC development through many financial and non-financial schemes. So far, the Singapore VC industry has grown into the third largest in Asia (AVCJ, 2002). Towards the end of the year 2000, there were already 118 VC fund management groups/companies in Singapore, both international and home-grown. Currently, there are more than 600 VC-backed companies in Singapore, with more than 500 investment professionals. Table 9.3 shows the profile of the Singapore VC industry from 1998 to 2000.

Table 9.3 The profile of VC industry in Singapore

	1998	1999	2000
Number of VC funds/firms	64	85	118
Estimated number of VC professionals	331	453	521
Total capital under management (US$ million)	4958	7426	9197
New funds raised (US$ million)	585	2231	1800
Opening investment portfolio (US$ million)	1484	1827	2697
New investments during the year (US$ million)	387	878	1129
Follow-on investments (US$ million)	12	132	141
Total investments in the year (US$ million)	399	1010	1269
Total divestments in the year (US$ million)	56	141	374
Closing investment portfolio (US$ million)	1827	2697	3593

Source: AVCJ (2002).

Even with the impacts of the Asian financial crisis and the economic recession, the VC industry in Singapore developed steadily in the 1990s under government support. From 1991 to 2000, Singapore experienced an average annual increase of about 31 per cent in total VC under management (see Figure 9.1).

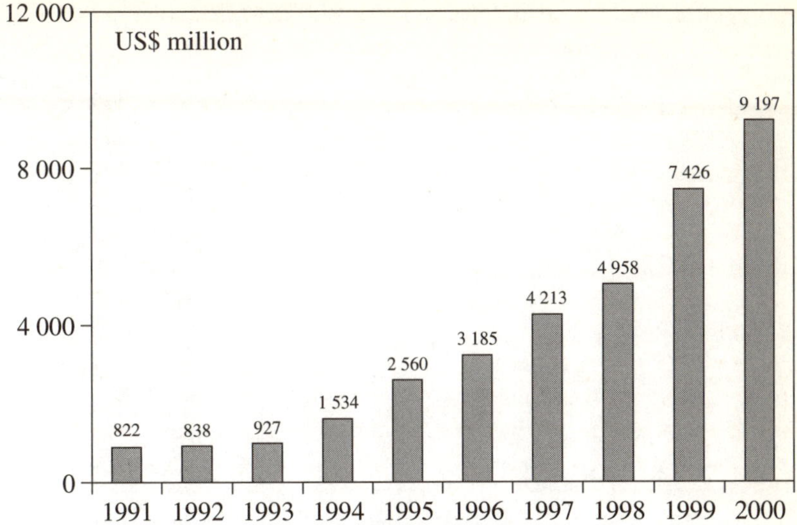

Source: AVCJ (2002).

Figure 9.1 VC pool in Singapore, 1991–2000

As far as the annual VC investment is concerned, it experienced the same development trend as total VC pools in the 1990s. In 1992, the annual VC investment in Singapore was just over US$100 million. However, by the end of the decade, the actual VC investment per year in Singapore had exceeded US$1000 million, an amount more than ten times that in 1992 (see Figure 9.2). Therefore the VC investment portfolio in Singapore also increased significantly in the 1990s (see Figure 9.3). On average, the VC investment portfolio in Singapore increased by about 28 per cent annually from 1991 to 2000.

With respect to the annual new VC funds raised, there were two booming periods in the development of the VC industry in Singapore (as shown in Figure 9.4). The first booming period was from 1993 to 1995. In fact, the new funds raised in 1995 were about 20 times more than the total raised in 1992. Another booming period started in 1999, just after the recession caused by the Asian financial crisis. The year 1999 also saw the highest annual new funds raised in the whole history of the VC industry in Singapore. Owing to the government's promotion of high-technology industries such as information technology and biotechnology in recent years, this booming period in VC investment is expected to last for some time, even though the trend has been declining since 1999.

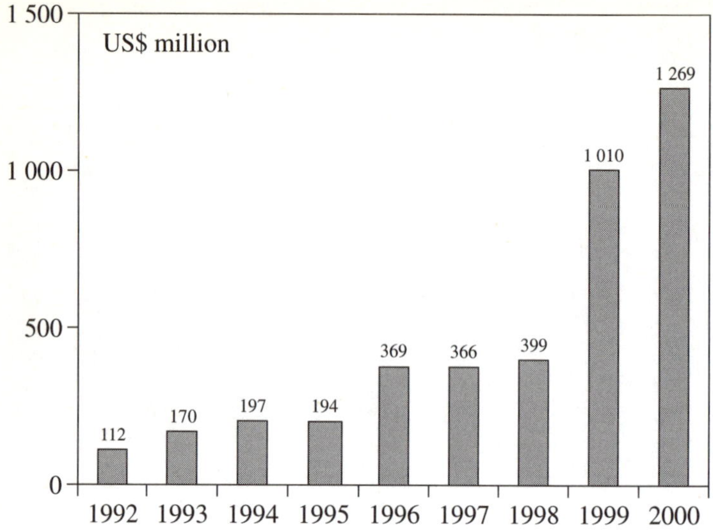

Source: AVCJ (2002).

Figure 9.2 Actual VC investment per year in Singapore, 1992–2000

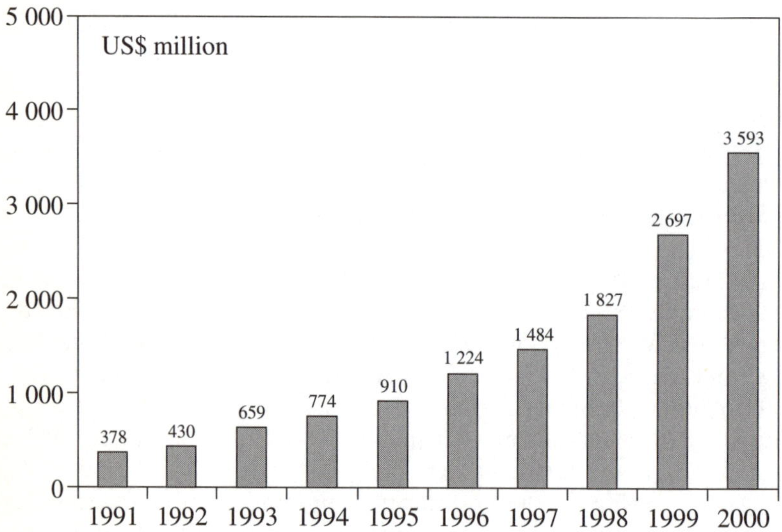

Source: AVCJ (2002).

Figure 9.3 VC investment portfolio in Singapore, 1991–2000

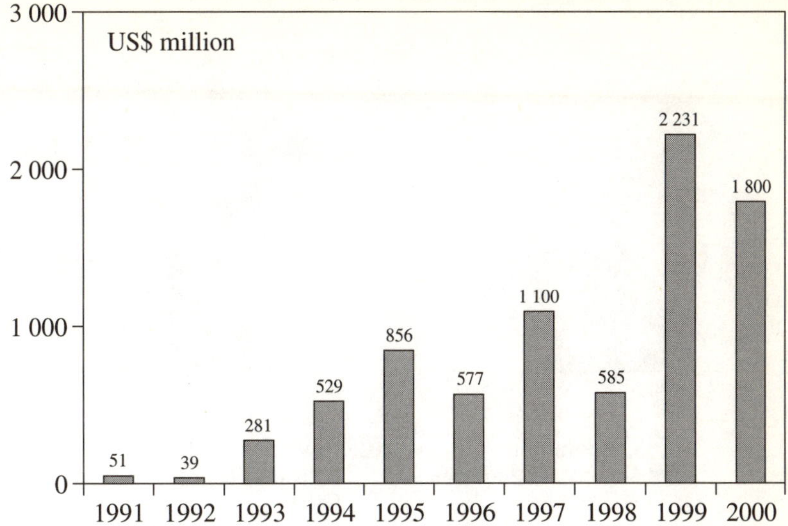

Source: AVCJ (2002).

Figure 9.4 Trends of new funds raised in Singapore, 1991–2000

Source of VC Funds in Singapore

As far as the sources of VC funds are concerned, corporations, government agencies and banks are the three major sources in Singapore. In 1999 and 2000, these three major sources accounted for around 75 per cent of the total new funds raised annually (see Figure 9.5).

As the regional financial hub, and in contrast to other Asian countries, Singapore has also attracted many VC investors from other countries. In recent years, more than two-thirds of VC funds in Singapore were from overseas. Geographically speaking, latest figures show other Asian countries and non-Asian countries contributing 31 per cent and 39 per cent, respectively to the VC investments in Singapore (see Figure 9.6).

Types of VC Firms

The Singapore VC market is characterized by a distribution of different types of venture capital firms. There are 'American-style' IVCs (independent limited partnership VC firms), 'Japanese style' FVCs (finance-associated VC firms) and CVCs (corporate-affiliated VC firms) in Singapore. This diversified VC market has been nurtured, to a large extent, by the Singapore government's

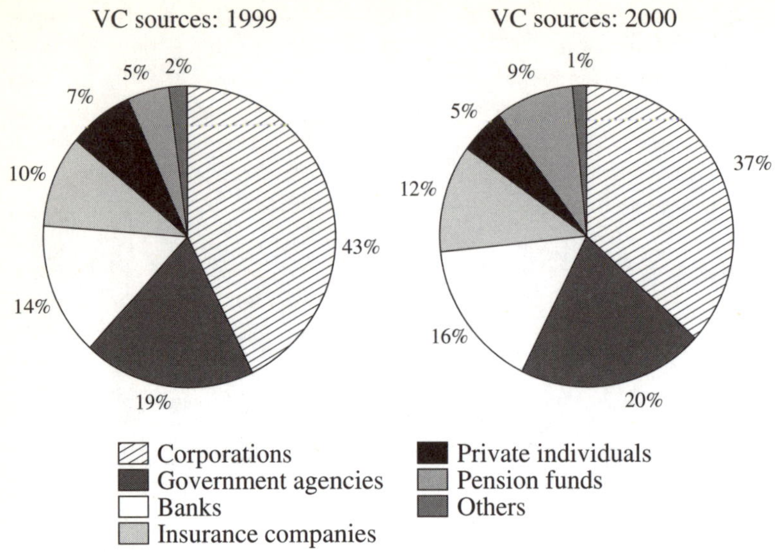

Source: AVCJ (2002).

Figure 9.5 Sources of venture capital in Singapore

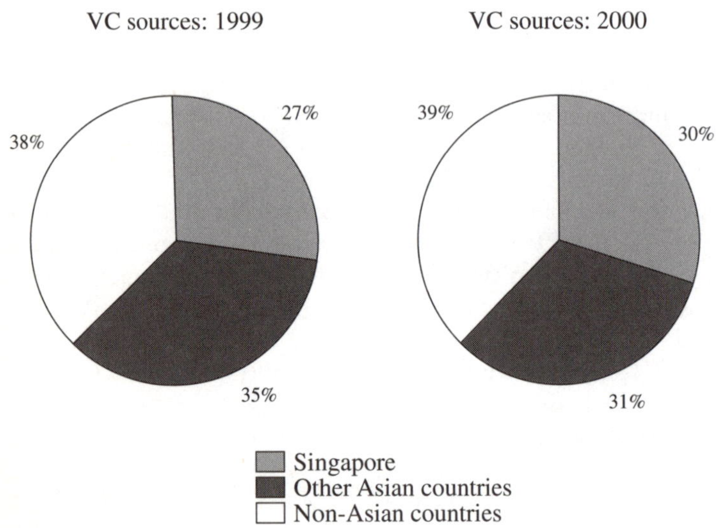

Source: AVCJ (2002).

Figure 9.6 Geographical breakdown of VC sources in Singapore (per cent)

open door policy, which has been implemented by EDB in promoting VC development in Singapore. Thanks to the beneficial government policies and strategic programs, cash-rich large corporations and government boards, as well as high net worth individuals and families, have established many strategic VC funds in Singapore. Thus the Singapore VC market consists of a mixture of various types of VC funding, that is business angels, independent VCs and finance-affiliated VCs, as well as government and corporate VC firms.

VC Investment in Singapore

Being an active regional financial hub, Singapore has not only absorbed overseas VC funds but also disbursed its funds to other countries. In recent years, funds invested locally in Singapore have accounted for less than 20 per cent of total VC disbursements. Nearly 60 per cent of the VC in Singapore has been invested in other Asian countries, which shows the importance of Singapore as a regional VC center. In addition, around 20 per cent of the VC disbursements in Singapore were invested in non-Asian countries in the last few years (see Figure 9.7).

In any investment decision process, the market segmentation of VC firms in stages and technology intensity are among the most important considerations, because stage and technology preferences are determined by the

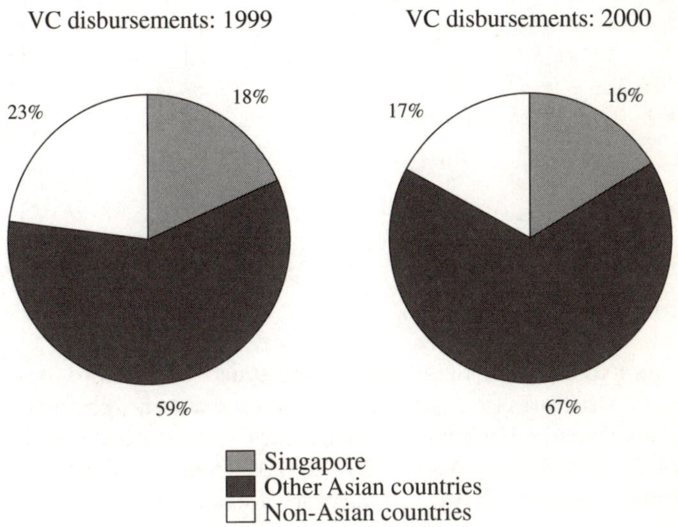

Source: AVCJ (2002).

Figure 9.7 Disbursements of VC in Singapore to companies, by region

level of risk tolerance of VC investors, which in turn reflects the entrepreneurship environment in a given country (Ruhnka and Young, 1991; Mayer, 2002).

Based on VC disbursements by financing stage, the entrepreneurship environment in Singapore reflects its tendency to be conservative. According to financing phases in recent years, about 70 per cent of the VC investments in Singapore flowed into expansion or later stages annually. In contrast, only about 30 per cent of the total VC investments were early-stage investments. For example, in 2000, only US$143 million VC funds, accounting for only 4 per cent of annual VC investments, flowed into seed-stage ventures. VC funds invested in the start-up stage amounted to US$934 million, or 26 per cent of the total annual VC investments. However, in the same year, nearly half (44 per cent) of the annual VC investments flowed into expansion-stage ventures (see Table 9.4). The contrast shows the characteristics of the entrepreneurship environment in Singapore: small risk-tolerance scales and conservative risk preferences.

Table 9.4 VC disbursements in Singapore by financing stage

	1999		2000	
	Per cent	Amount (US$m.)	Per cent	Amount (US$m.)
Early stage	28	755	30	1078
Seed	6	162	4	144
Start-up	22	593	26	934
Expansion	49	1321	44	1581
Mezzanine	14	378	10	359
Buy-out	5	135	12	431
Turnaround	4	108	4	144
Total	100	2697	100	3593

Source: AVCJ (2002).

According to the EDB, Singapore has expanded rapidly with the change in industrial structures from being labor-intensive sectors to being technology-intensive sectors. High-tech industries have grown into one of the main sources of GDP development (EDB, 2002). Therefore the technology-focused disbursements of VC in Singapore are consistent with the industrial structure characteristics.

As shown in Table 9.5, high-tech industries usually attract nearly two-thirds of the annual VC investments in Singapore. Take the year 2000 as an example. Computer products/services, information technology, heavy man-

Table 9.5 Disbursements of VC in Singapore, by technology

Industry	1999 Amount (US$m.)	1999 Per cent	2000 Amount (US$m.)	2000 Per cent
High-tech industries	1774	65.8	2283	63.5
Computer-related	226	8.4	277	7.7
Electronics	391	14.5	407	11.3
Information technology	405	15.0	563	15.7
Manufacturing – heavy	51	1.9	51	1.4
Medical/biotechnology	275	10.2	325	9.0
Telecommunications	426	15.8	661	18.4
Other industries	923	34.2	1310	36.5
Agriculture/fisheries	5	0.2	6	0.2
Conglomerates	39	1.5	56	1.6
Construction	20	0.7	200	5.6
Consumer production/services	326	12.1	346	9.6
Ecology	20	0.7	25	0.7
Financial services	157	5.8	173	4.8
Infrastructure	62	2.3	79	2.2
Leisure/entertainment	38	1.4	50	1.4
Manufacturing – light	55	2.1	71	2.0
Media	5	0.2	18	0.5
Mining and metals	12	0.4	113	3.1
Retail/wholesale	9	0.3	9	0.2
Service – non-financial	46	1.7	32	0.9
Textiles and clothing	19	0.7	19	0.5
Transportation/distribution	33	1.2	21	0.6
Travel/hospitality	21	0.8	32	0.9
Utilities	56	2.1	59	1.6
Total	2697	100	3593	100

Source: AVCJ (2002).

ufacturing, medical/biotechnology and telecommunications attracted US$2283 million in VC funds, accounting for nearly 64 per cent of total VC investments. Among these high-tech industries, telecommunications, information technology and electronics have been the most attractive to VC investors. In total, the VC disbursements to these three industries have accounted for nearly 50 per cent of the annual VC investments in recent years.

Aiming to develop Singapore into a biomedical sciences hub with world-class capabilities across the whole value chain, the Singapore government has adopted many integrated approaches to promote the development of the medical/biotechnology industry since the late 1990s. The EDB launched a biomedical sciences program in 2001, setting up a Biomedical Sciences Investment Fund and other related funds. These VC funds, focusing specifically on the medical/biotechnology industry, have committed some S$130 million to 14 projects in Singapore and overseas ventures/projects (EDB, 2002). In addition, the VC disbursements to the medical/biotechnology industry have increased rapidly in recent years. Figures for 1999 and 2000 show that the medical/biotechnology industry has already been attracting about 10 per cent of VC investments annually.

Along with the manufacturing industry, the service industry constitutes the basic industrial component of the Singapore economy. Hence VC investors in Singapore have also paid a great deal of attention to the service industry, such as consumer production/services and financial services. In fact, among all the non-high-tech industries, consumer production/services and financial services attract the highest amount of VC investments. In recent years, nearly 20 per cent of all the VC investments flowed into these services.

VALUE-ADDED ACTIVITIES BY VC FIRMS IN SINGAPORE

Past studies have shown that VC firms provide not only money but also value-added activities (for example Rind, 1981; Tyebjee and Bruno, 1984; Jain and Kini, 1995). Investee firms could receive benefits such as sound management advice, formulation of market strategy, technology and capital market access as well as other networking connections. Even in pre- and post-IPOs, investee firms enjoy the prestige that comes with being certified by VC firms compared to firms of non-VC backed IPOs which have no such certification (Megginson and Weiss, 1991).

To find out the characteristics of value-added activities in the Singapore market, we examine the value-added support from the VC perspective. By taking such an approach we can gain access to a wide range and a large portfolio of investees. We carried out case studies involving ten randomly chosen VC firms in Singapore to highlight the value-added support provided to their investee companies. The profiles of the ten VC firms are shown in Table 9.6.

This sample is a good representation of venture capital firms in the Singapore market. It covers VC firms founded at different times over the past two decades with five foreign and five domestic firms, which reflects the local government's open policies in developing venture capital. The limitation of the sample size is partially offset by the demographic distribution in

Table 9.6 Profiles of ten VC firms in Singapore

Profiles/companies	Case A	Case B	Case C	Case D	Case E	Case F	Case G	Case H	Case I	Case J
Year founded	1995	1997	1986	1997	2000	1990	2000	1984	1991	1997
Nationality	F	D	D	D	F	F	F	D	D	F
Number of founders	3	3	2	2	2	—	—	—	3	—
Number of professionals	17	17	22	2	2	17	2	9	6	15
Type of firm	JV	IVC	IVC	IVC	IVC	IVC	CVC	IVC	IVC	IVC
Number of portfolio companies	>100	35	100	16	5	112	11	>100	56	21
Number of funds raised	3	—	3	1	—	13	1	—	2	1
Total funds raised (S$ million)	710	>100	1 300	20	250	405	75	160	180	400
Funds invested(S$ million)	550	—	—	—	—	400	—	—	—	—
Types of financing	Mi	Mi	E/Ma	E/Mi	E/Ma	Lo/Ma	Ma	Lo/Ma	E	E/Mi
Financial role	L/S	S	L	L	S	S	L/CL	L	—	L
Preferred stage of investment	E/LS	ES	LS	E/LS	LS	E/LS	ES	LS	E/LS	E/LS
Geographic preference	A/Na	A	A	A/Na	A	A/Na	A/Na	A	A/Na	A/Na
Preferred investment size (S$ million)	2–10	2–100	>40	1–5	1–5	0.5–5	2–5	10–50	1–10	5–25
Industrial preference	Elec, IT	High-tech	High-tech	High-tech	High-tech	High-tech	High-tech	High-tech	High-tech	High-tech

Notes: F = foreign, D = domestic, JV = joint venture, IVC = independent VC, CVC = Corporate VC, E/Mi = equity/minority equity, E/Ma = equity/majority equity, Lo = loans, L = lead investor, CL = co-lead investor, S = syndicate investor, ES = early stage (start-up, early development), LS = later stage (expansion, pre-IPOs), A = Asia region, Na = non-Asia region, Elec = electronics, IT = IT sector, High-tech = high-tech sectors, Ntech = non-high-tech sectors.

terms of firm and portfolio sizes, types of organizations, different preferences in stages and business and technology sectors, and different roles of venture capitalists in investing (such as syndication).

We group the value-added programs and tools into four main categories: technological, marketing, management and financial supports. Using the comments given by managers/partners from the VC firms on their portfolio management practice and value-added activities, we compiled the following criteria for assessing the strengths of various value-added activities:

1. To stipulate the requirement for 'board seats' or act as 'lead investors' is seen as a signal of participation in leadership formulation and control.
2. Regular financial reporting enhances the VC firm's role in financial control and strategic decisions.
3. The presence of a board of technology advisers or forging intensive relationships with scientific or technical institutions provides technology access and research assistance.
4. Previous VC experiences of IPOs or M&As are strong indicators of the ability to provide financial support and networks.
5. VC firms with global offices provide a solid base for international market penetration.
6. Corporate VC firms can rely on their parent companies with deep industrial knowledge to provide marketing support, both in strategy formulation and in implementation.

Using such criteria, we summarize in Table 9.7 the value-added activities of the ten VC firms in Singapore.

Description of Findings

By interviewing ten VC firms with different backgrounds and characteristics, we found that the VC firms launch different programs and use various kinds of portfolio management tools aimed at increasing the odds of success of investee firms. After comparing the programs and tools, we found some activities are prevalent in all ten VC firms, as well as other activities which are unique to individual VC firms. Furthermore we observed that the strategies of value-added programs and tools are related to the backgrounds of the VC firms. These value-added activities depend, to some extent, on the investment strategies of VC firms in terms of stage and technology preferences and the roles of VC firms as lead investors, co-lead investors or non-lead investors.

Table 9.7 Value-added activities provided by VC firms in Singapore

Categories/companies	Case A	Case B	Case C	Case D	Case E	Case F	Case G	Case H	Case I	Case J
Planning finance/strategies	++	++	++	+	++	++	++	++	++	++
Controlling finance	++	–	++	+	+	+	++	+	+	–
Controlling additional investment(s)	++	–	++	+	+	+	++	–	–	–
Controlling leadership	+	+	++	+	++	++	++	++	+	++
Following production/research	–	–	+	–	+	+	+	+	–	–
Helping to get additional financial resources/bank loans	++	+	++	++	++	+	–	++	+	+
Participating in leadership formulation	+	+	++	+	++	++	++	++	+	++
Helping to recruit key personnel	–	–	+	–	–	–	–	+	–	+
Helping to get/access new technology	++	+	+	+	+	++	++	+	+	+
Helping in connection/network building	++	+	++	+	++	++	++	++	+	++
Helping to expand to the international market	+	+	++	+	++	++	++	++	++	++
Participating in marketing strategy formulation	++	++	++	+	++	++	++	++	+	++

Notes: The strength of the connection is noted by '+' and '–' signs: strong participation/help is indicated by '++'; occasional (semi-strong) participation/help is indicated by '+'; no participation/help is indicated by '–'.

Most Commonly Used Value-added Activities

'Planning finance/strategies' is the most common value-added activity offered by VC firms because it provides the guidelines for both venture capitalists and entrepreneurs to realize the potential growth of the ventures after the injection of VC funds. The second most common support provided by VC firms is 'Participating in marketing strategy formulation', followed by 'Helping in connection/network building' and 'Helping to expand to the international market'. These active roles in global networking exhibit the internationalization of VC sources and disbursements in Singapore, thus demonstrating the 'global-oriented' economy of Singapore. Moreover this reflects the fact that venture capitalists emphasize the importance of marketing strategies. In other words, marketing risk is, at least, one of the greatest concerns before and after making investments. We also observed that getting control of the investee firm is one of the major concerns of VC firms. They can gain the control rights by sitting on the board of directors or requiring the right to replace the founders, as was the situation in case C.

'Helping to get additional financial resources/bank loans' and 'Helping to get/access new technology' are other value-added support services provided by the VC firms. This reflects the relative maturity of the Singapore VC market regarding VCs' intermediary roles. In particular, VC firms with very strong financial/bank backgrounds usually provide the support of 'Helping to get additional financial resources/bank loans'. For example, one of the sponsors of case I is a large commercial bank in Singapore. Such VC firms are more likely to provide valuable financial resources to assist investees in the expansion of their businesses. On the other hand, corporate VCs, for example in case G, often make use of the technological advantages of their parent companies to provide investees with new technological support.

Least Often Used Value-added Activities

The least often used value-added activity in the Singapore market is 'Helping to recruit key personnel'. According to the results of our interviews, most VC professionals prefer not to recruit key personnel for investee companies because this may result in conflict between the founders (entrepreneurs) and outside professional managers, as the latter may have different, or even sometimes opposite, philosophies of risk taking, and thus different growth strategies. In addition, the relatively weaker participation of VC firms in 'Following production/research' could be attributed to their emphasis on pre-investment screening strategies and their confidence in investee firms' management teams.

Characteristics of VC Firms and Corresponding Activities in Adding Value

In terms of 'Controlling finance' and 'Controlling additional investments', the VC firms show a wide range of strategies. For example, the VC firms in case B and case J seldom participate in 'Controlling finance' and 'Controlling additional investments' while those in case A, case C and case G have strong participation in such activities. Some degree of difference is also found in VCs' participation in 'Controlling leadership' and 'Participating in leadership formulation'. Furthermore, differences exist in VC firms' activities in helping to gain access to new technology and capital markets. Such differences are related to unique characteristics (for example, background of VC investors) of the various VC firms and their investment strategies (stage and technology preference), as well as their status in the investee firms (for example, as lead investor, co-lead investor or syndicate investor).

A comparative analysis of the value-added activities of VC firms reveals four kinds of corelationships between the investment characteristics of venture capital firms and their value-added activities.

1. Parent companies with deep industry roots have an advantage in providing value-added support to investee firms (technological support, customer connecting and market penetration), as found in case G. Furthermore, such VC firms would be more capable of executing M&As within the industries, which would facilitate the acquisition of new technology and cooperation among the incumbents in the industry.
2. The differences in the ability to provide networking with capital markets/banks and technologies are corelated to the different backgrounds and experiences of VC firms. Usually financial affiliated VC firms have more financial resources to help investee firms get into the international capital market and to gain access to other investors' resources (as in cases C, D, E and H), while industrial corporations or government technology agencies can help investee firms in technology cooperation, as in cases F and G.
3. VC firms focused on early-stage investments tend to prefer to participate in business-level activities, as found in case B. Such a high level of participation is essential to new ventures, especially those founded by those with technological expertise. The involvement may help the investee companies to re-examine their business models, to determine their financial/strategic planning, and to expand regional and global markets.
4. Lead investor VC firms have stronger control of investee firms because of their status as a majority shareholder. Usually, they win the right to replace the founders of investee firms, as was found in case C. Moreover, they have veto rights in major financial and business decisions. They

may even have influence in financial and operational management activities such as the recruitment of key personnel. In contrast, non-lead investor VC firms may face difficulties in getting the investee management team to adopt their strategies owing to their status as minority shareholders, as was observed in case E.

CONCLUSIONS

For nearly 20 years, the contributions of an active venture capital market in Singapore have been significant in developing the country's economy and sustaining industry competitiveness. The Singapore VC market has encouraged waves of technological innovations as well as incubating new, knowledge-intensive venture creations. The Singapore VC market is relatively large and has industry breadth among emerging markets in Asia. In terms of size, Singapore's VC pool is the third largest in Asia, next only to Hong Kong and Japan (AVCJ, 2002). In terms of industry breadth, the sector distribution of VC-backed companies in Singapore is broader, with a higher concentration of IT, biotechnology and telecom sectors compared with other Asian countries (ibid.).

In addition, the Singapore VC market enjoys a fairly balanced distribution of different structures of VC firm types. There are 'American-style' IVCs (independent limited partnership VC firms) as well as 'Japanese-style' FVCs (finance-associated VC firms) and CVCs (corporate-affiliated VC firms) in Singapore. This diversified VC market has been nurtured by the Singapore government's open door policy, which has been largely implemented by the Singapore Economic Development Board (EDB), in playing an important role in promoting VC development in Singapore. We found that independent VC firms in Singapore prefer investments in the expansion and later-stage development, whereas corporate or government associated VC firms exhibit preferences for early-stage, high-technology ventures (Wang *et al.*, 2002).

Furthermore we can see that venture capital firms in the Singapore market play an active role in providing value-added support services to investee firms. Besides succeeding in providing financial capital, they serve well as intermediaries in dealing with the information asymmetries between the investors and entrepreneurs and between the entrepreneurs and various markets. Their activities cover a wide range of supports, including financial and business strategy formulation, controlling and monitoring, and providing market and technology assistance. In particular, the global market focus of venture capital firms in Singapore demonstrates the continuing economic trends of the Singapore market.

Although both foreign firms and local firms in the Singapore market tend to concentrate their investment focus on Asia and the Pacific Rim region, and to provide strategic and international marketing value-added support, they also show different styles in some aspects of monitoring and adding value to investee firms. The VC firms show a wide spread of strategies in 'Controlling finance' and 'Controlling additional investments'. We also found some differences in VC firms' participation in 'Controlling leadership' and 'Participating in leadership formulation'. Furthermore, differences exist in VC firms' helping to gain access to new technology and the international capital market.

We speculate that there are two possible factors contributing to such differences in value-added activities. One is the unique characteristics of VC firms (such as the background of VC investors). The other is the investment strategies of the VC firms (stage and technology preferences) and their status in investee firms, (for example, as leader investor, co-leader investor or syndicate investor). Different types of VC firms adopt different roles in market segments (different stages and technology intensity) and provide different value-added activities. Our case studies also show that there is a fundamental difference between these venture capitalists, based on investment preferences in terms of stages of investment and technology intensity. In other words, different forms of venture capital firms have fundamentally different investment preferences in stages and technology intensity, and thus play different roles in promoting venture growth in specific stages and technologies (Mani and Bartzokas, 2002).

According to our interviews and analysis, among the critical factors the governance structure may be the most critical to the risk tolerance levels of VC firms, which in turn determine their risk preference (Wang and Zhuang, 2002). Within our analysis framework, venture capitalists could be categorized into several groups: business angels, independent limited partnership venture capital firms, and affiliated venture capital firms (both financial affiliated VCs and industrial affiliated VCs). We differentiate the three groups of venture capitalists with different governance structures: individual, limited partnership and hierarchy, respectively. The basic claim in our analysis is that the different governance structures influence the variables of investors' risk tolerance scales, which in turn determine their investment preference in stages or business sectors.

Using the theoretical hypotheses of a venture's 'stage of development', we conceptualize the investments in different stages and technology intensity as 'assets' with different risk–return profiles (Wang and Zhuang, 2002). Thus the investment choice in stages and technology intensiveness corresponds to investors with different risk tolerances and different capabilities of dealing with associated risks. Consequently VC firms with different governance structures should exhibit some differences in such investment strategies as stage breakdown and technology preference, that is, investments characterized by different risk–return profiles. As a result they provide different value-added support which is associated with the development stages of ventures.

The Singapore VC market with its different organizational structures (government, public and private) contributes actively to regional technological and economic development. Also the heterogeneous characteristics of VC firms and their value-added activities for investee firms imply that, when launching initiatives or support programs, decision makers need to realize the impacts of governance structures on their investment strategies, especially in stage and technology choice. This is in contrast to taking VC firms as a homogeneous group. Different investment preferences would determine the scope of their value-added activities and therefore have different impacts, as a whole, on economic development. The government and its agencies should encourage venture capital providers to fine-tune their governance structure and consider changing incentive arrangements by providing for legal frameworks or economic incentives to achieve long-term objectives.

Furthermore, as Singapore is a relatively large VC presence in emerging markets, its success could have implications in the institutional structure of the VC market for other developing countries. When financial gaps in innovative start-ups become problem issues, business angels and corporate venture capitalists could be invited to inject more money in the early stages in fostering technological innovations, given that many corporate VCs are strategic investors. 'Safety-net' subsidiary programs and a discriminatory tax policy favoring the business angels could also be enhanced. If encouraged to fine-tune their governance structure, independent VC firms might change their risk preference. Some kind of compensation policy for loss in early-stage investments may encourage IVCs to invest in more technology-intensive venture creations.

NOTE

1. Singapore Exchange (SGX) is Asia-Pacific's first demutualised and integrated securities and derivatives exchange. SGX was inaugurated on 1 December 1999, following the merger of two established and well-respected financial institutions – the Stock Exchange of Singapore (SES) and the Singapore International Monetary Exchange (SIMEX). Singapore Exchange maintains two separate listings for equities. The mainboard is made up of more than two-thirds of the companies listed on Singapore Exchange. A second board, SESDAQ, was established to provide an alternative avenue for small and medium sized companies to raise funds from the stock market.

REFERENCES

Asian Venture Capital Journal (AVCJ) (2002), *The 2002 Guide to Venture Capital in Asia*, Hong Kong: Asian Venture Capital Journal.
Buchanan, W.T. (1989), 'Venture funds: a special survey', *Singapore Business*, January, 14–15.
Chia, K.G. and K.C. Wong (1989), *Venture Capital in the Asia Pacific Region*, Singapore: Toppan.
Chin, C.Y. (1988), *Venture Capital Business in Singapore*. Singapore: DBS Bank, Economic Research Department.
EDB (various years), *EDB Annual Report,* Singapore: Economic Development Board.
ERC (2002), *Report of the Entrepreneurship and Internationalisation Subcommittee*, Singapore: Economic Review Committee.
Jain, B.A. and O. Kini (1995), 'Venture capitalist participation and the post-issue operating performance of IPO firms', *Managerial and Decision Economics*, **16**(6), 593–606.
Mani, S. and A. Bartzokas (2002), 'Institutional support for investment in new technologies: the role of venture capital institutions in developing countries', *Intech*, The United Nations University.
Mayer, C. (2002), 'Financing the new economy: financial institutions and corporate governance', *Information Economics and Policy*, **14**(2), 311–26.
Megginson, W. and K. Weiss (1991), 'Venture capitalist certification in initial public offerings', *Journal of Finance*, **46**, 879–903.
Rind, K.W. (1981), 'The role of venture capital in corporate development', *Strategic Management Journal*, **2** (2), 169–80.
Ruhnka, J.C. and J.E. Young (1991), 'Some hypotheses about risk in venture capital investing', *Journal of Business Venturing*, **6**, 115–33.
SGX (2002), *SGX Annual Reports*, Singapore: Singapore Exchange.
SVCA (2002), *Singapore Venture Capital & Private Equity 2002*, Singapore: Singapore Venture Capital Association.
Tyebjee, T.T. and A.V. Bruno (1984), 'A model of venture capitalist investment activity', *Management Science*, **30**(9), 1051–66.
Wang, C.K. and W.T. Zhuang (2002), 'Differences in stage, technology, risk preference and governance structure of venture capital', working paper, National University of Singapore.
Wang, C.K., N. Wrede and Q. Lu (2002), 'Investment decision process and type of venture capital firms', working paper, National University of Singapore.

10. High-tech venture capital investment in a small transition country: the case of Hungary

László Szerb and Attila Varga

INTRODUCTION

Hungary started to change its economic system from a planned economy to a market economy in the late 1980s. In the first years of transition, GDP declined by around 20 per cent, unemployment increased from zero to 15 per cent, and inflation began to rise. At the same time economic restructuring started and liberalization also helped to encourage entrepreneurship. Since the beginning of the transition, one of the major problems of small and start-up businesses has been the lack of financial resources. While the government promoted new business creation in the early years of transformation, this support had disappeared by 1995. At the same time, the financial system of Hungary began to improve by developing banking and credit institutions and stock markets.

Since 1997 there has been a stable positive growth of the country amounting to a 5.2 per cent growth rate in 2000. This development was mainly due to large foreign companies well equipped with capital and management knowledge. At the same time, small business growth showed a declining trend that was turned around only in 1999–2000. Small business debt financing has been improving for the last two years, however, equity financing is still a problem.

As one of the solutions to equity financing, venture capital appeared first in the country in the early 1990s, but played a minor role up to 1995. Despite higher than average political and economic risk, the small size of the country and limited income, the Hungarian venture capital market went through major changes and development, following worldwide tendencies. By 2001, venture capital had become an important player in early-stage high-tech investment.

This chapter has the following structure. In the next section the historical evolution of the venture capital industry in Hungary is presented and the role of government and other agencies in initiating the venture capital industry is also covered. The third section highlights the current structure of the venture capital industry in Hungary, focusing mainly on high-technology development. The fourth section presents a case study on high tech projects financed by five venture capital companies, with detailed descriptions of venture capital companies and investee firms. Interactions between the venture capital companies and the investee firms, including the value-added support, are also analysed. We conclude with a critical evaluation of the main findings of the study.

HISTORICAL DEVELOPMENT AND THE BACKGROUND OF THE VENTURE CAPITAL INDUSTRY IN HUNGARY

In Hungary, venture capital emerged at the end of the late 1980s, but did not play a major role until the mid-1990s. Initially the Hungarian government was actively involved in the establishment of classical venture capital companies such as Innofinance, Covent and Multinova. The first real venture capital fund, First Hungary Fund, was founded in 1989. Between 1989 and 1992, the venture capital market developed dynamically, with the Hungary only funds dominating. As a sign of the development of the venture capital market, the Hungarian Venture Capital Association was established in 1991.

In these early years of transition the venture capital market was very small and associated with a high political and economic risk. However, at the same time, transformation into a market economy was starting: market institutions were established, trade prices and the start-up of new ventures became free, and the private sector share of production began to rise. By 1992, more than one-third of companies based on the subscribed capital were privatized and, by 1993, more than half of the gross domestic profit (GDP) was produced by the private sector. In part the last old regime, but mainly the democratically elected Parliament of 1990, created new laws consistent with the needs of a market economy. We have to emphasize here the Company Act (1998), the Transformation Act (1989), the Bankruptcy Act (1990), the Competition Act (1990) and the Security Act (1996) in this respect (Szerb and Ulbert, 2002).

In the early 1990s, as a sign of restructuring, over two million jobs were lost, mainly in the large business sector, owing to 'downsizing' and privatization. At the same time, microenterprises and small firms increased employment by creating hundreds of thousands of new jobs. Between 1990 and 1995, the number of businesses almost tripled, resulting in over one million independent business entities.

However, the further development of the domestic private sector and especially that of the large number of small and medium-size enterprises

(SME) was not satisfactory. The lack of proper financial resources (both equity and credit) proved to be a major problem mainly in the case of the most prosperous entrepreneurial businesses (Kállay, 2000). For a few years it was a common belief that venture capital as one of the most commonly used financial methods would help to solve small businesses' growth problems. Probably this was the reason of the establishment of two venture investment companies by the state in the early 1990s. However, owing to circumstances such as undercapitalization of funds as well as bad investment strategy towards high-risk and low-return projects, the companies had to give up financing small business (Karsai, 1997).

Favorably changing political and juridical conditions connected with the needs of restructuring, management knowledge and the lack of capital opened new perspectives for venture capitalist. In 1995, coupled with the renewal of privatization, a new wave of venture capital investment led by foreign funds started. Table 10.1 summarizes the most important characteristics of the venture capital industry since 1995. We should note, however, that the data are not perfectly reliable, since internationally accepted data collection was started in Hungary by the European Venture Capital Association only in 2000 (Karsai, 2002)

Table 10.1 Trends in venture capital investment in Hungary, 1995–2002 (data in million US$ if not stated otherwise)

	1995	1996	1997	1998	1999	2000	2001
Estimated amount of VC pool	NA	NA	150	230	350–400	400–500	440–500
Cumulative funds raised in Hungary	590	740	1000	1260	1380	NA	NA
New funds raised	60	140	280	160	100	68	56
New investment in year	90		150	60	41	103	64
Total invested capital	500	590	740	780	820	925	1000
Average investment	~2–2.5	~2–2.5	~5	5	3.4	2.2	2.3

Note: The 90 investment applies to 1995 and 1996 jointly.

Sources: *Hungarian Venture Capital Association Yearbook* (1999, 2000, 2001, 2002), Karsai (1999), and own primary data collection.

By 1996, the total investment of the 25 members of the Hungarian Venture Capital Association (HVCA) exceeded US$250 million (US$590 million total), but only 52 per cent of the available capital was used. The most notable year was 1997, when the highest amount, US$150 million was invested. According to the HVCA, by 1998 the available venture capital was about US$1.2 billion and out of this US$780 million was effectively invested. The estimated amount of the venture capital pool increased dynamically to around US$500 million per year, including the regional funds' estimated share devoted to Hungary. Similar to the VC pool, cumulative funds and new funds showed high growth in the second part of the 1990s, however, the trend was broken by the 1998 Russian crisis.

Increased activity of foreign investors characterized the second part of the 1990s: in 1995, domestic venture capital funds amounted to 70 per cent of VC investment. By 1998, foreign venture capitalists supplied about US$350 million (46 per cent) of venture capital investment, and the Hungarian state supplied about one third of it. Domestic private investors' share was less than 25 per cent of total investment (*Hungarian Venture Capital Association Yearbook*, 1999). We have no reliable data regarding the national origin of the venture capital, but it follows the international tendencies, mainly with Anglo-Saxon-country dominance. Up to 1998, American institutional investors provided about two-thirds of the foreign venture capital investment, while the share of European institutional investors, including pension funds and banks, was around one-third (ibid.).

Since 1999, the domestic proportion of transactions has been stabilizing at around half of the deals, but less as regards the value of investments (*Hungarian Venture Capital Association Yearbook*, 2001). Meanwhile international deals have an increasing share, about one-third of the total transactions. At the same time, Hungarian venture capital funds (such as Corvinus) are more interested in foreign investments, mainly in neighboring countries like Croatia, Romania, Slovakia and Ukraine (see the next section).

The amount of the invested venture capital in Hungary is not known exactly. Halaska and Kovács (1999) reported that venture capitalists invested US$300–500 million in Hungarian businesses between 1995 and 1999, with the increased involvement of foreign regional funds. Karsai and Rácz (2000) claimed that, by 2000, the invested venture capital exceeded US$1 billion. Another study (B.I., 2001) stated that total venture capital investment was about US$800 million by the end of 2000 and this is about the same amount that the HVCA estimates (*Hungarian Venture Capital Association Yearbook*, 2000). If we add the US$64 million investment that was made in 2001 (*Hungarian Venture Capital Association Yearbook*, 2001), the estimated total amount of venture capital investment was about US$860–1000 million by the end of 2001. A reason for the differences amongst the estimates may be the mixing of private equity and venture capital investments. In direct equity investment, the European Bank of Research and Development (EBRD) has played a major role. Note that, in Table 10.1, EBRD investments are also included. If we accept that venture capital investment is around US$1 billion, it constituted around 5 per cent of the total foreign direct investment (FDI) in Hungary by year 2001.

There is an even less reliable estimate that we can make regarding the number of companies receiving venture capital injection. Over the years 1999–2001, there were 87 deals, but many of them were second or third-round investments in the same company. Judging by personal interviews with venture capital experts, the number of companies receiving venture capital was about 200–450 up to 2002. We should note, again, that there are no exact data available regarding the above number.

Several studies (for example, Ludányi, 2001a; Karsai, 1999, 2000) have claimed that the Hungarian venture capital market was the most attractive amongst transitional countries in the second half of the 1990s. In 1996, the US$350 million venture capital investment reached 0.8 per cent of the Hungarian GDP, higher than in some developed or European Union countries. According to a recent investigation by the *Global Entrepreneurship Monitor* (GEM) (Reynolds *et al.*, 2001, Acs *et al.*, 2002), the average of the GEM-24 country venture capital investment was about 0.5 per cent of GDP, and was less than 0.1 per cent of GDP in the case of Hungary. Other estimates take this share about 0.5 per cent of the GDP (*Hungarian Venture Capital Yearbook*, 2002). If we compare the results from 1996 to 2001, we can conclude that, relative to the other parts of the world, the venture capital market in Hungary has declined. In other words, the renewal of the venture capital market, mainly in the USA in the late 1990s, had a limited effect on the Hungarian venture capital market. It can be seen that positive world market events affect Hungary with a certain lag: 2000 was an exceptionally good year in Hungary. However, bad events hit Hungary very quickly: the further development of the venture capital market was halted by the collapse of the major new economy companies at the end of 2000.

In the second half of the 1990s, there was a change in government involvement in the venture capital market. In 1997, the fully state-owned Hungarian Development Bank (HDB) was reorganized and became the most important source of direct state intervention in the venture capital industry, helping to restructure several companies. HDB also owns 11 regional development companies that finance smaller and early-phase businesses. The help and the financial scheme of these development companies are more similar to ordinary bank loans than to venture capital investments. Overall, the relative importance of the HDB and its regional development companies has been limited (Karsai, 2002).

In order to encourage new venture capital formation the Hungarian Parliament passed the Venture Capital Act (VCA, Act XXXIV of 1998), the first in the Central–Eastern European region. Domestic companies and funds registered under the VCA have to be licensed by the State Money and

Capital Market Commission (SMCMA). The amount of subscribed capital has to be over HUF500 million (around EURO200000), and the VCA describes the documentation that it is necessary to submit to the SMCMA, including the Deed of Foundation and the Rules of Organization and Operation. In exchange the VCA provides tax allowances and potential collection of the funds from the public.

Unfortunately, since its existence, there were only two companies receiving permission to register under the VCA, and at present none of them operates under the VCA. These data show that the law has had a marginal effect on the Hungarian capital market. The main reason for the failure of the VCA was associated with the strict rules on investment. Experts agreed that the VCA, as it stood, was overregulated, and the rigid requirements were not compensated for by the tax allowances (Karsai and Rácz, 2000; B.I., 2001).

Besides the regularly published yearbooks of the HVCA, there have been only a few major empirical studies analyzing the role and the profile of venture capital in Hungary: Karsai (1999), Lemák (2000) and Ludányi (2001a, 2001b). Before 1995, when the Hungarian state dominated in the venture capital industry, the main type of transaction was turnaround investment and management buy-outs (MBOs). As we noted earlier, after 1995, foreign investors took the leading role. Investigating the characteristics of venture capital investment in 1996–7, Karsai (1999) found that venture capital investments mostly concentrated on medium and large enterprises. The average amount of investment was about two times more than the European Union average (in 1996, ECU2.6 million to ECU1.01 million; in 1997, ECU5.07 million to ECU1.81 million). Venture capitalists decreased the high risk by investing in larger companies. Of the total investment, 90 per cent focused on the expansion phase (development capital). Turnaround businesses constituted half of the financing and seed capital was absolutely zero. Although MBO played an important role before 1995, it represented only 4 per cent of the investment in 1999. Lemák (2000) reported that foreign regional funds had become the main venture capital investors in Hungary by 2000.

International capital movements highly influenced Hungarian venture capital investment. Before 1998, venture capitalists were interested in services, machinery, food industry and information technology sectors. Since 1999, interest in high-tech enterprises, such as information technology, software firms, telecommunication firms associated with the Internet and biotechnology-type businesses has been rising.

THE STRUCTURE AND THE CHARACTERISTICS OF THE VENTURE CAPITAL INDUSTRY IN HUNGARY, 1999–2001

Following worldwide tendencies, the Hungarian venture capital market has gone through two major crises over the last few years. Table 10.2 presents the most important data regarding the 1999–2001 period. The 1998 Russian crisis hit the venture capital sector badly; there were only 12 transactions, with US$41 million of invested capital. However the damage in 2000–2001 was not as severe. The downturn in terms of both the number of transactions and the invested capital was about 40 per cent, but the early-stage technology investment showed noticeable resilience (*Hungarian Venture Capital Association Yearbook*, 2001). The average amount of investment of US$2.2–2.3 million was slightly higher than in the previous years (around US$2 million) in 2000–2001, though the typical investment was in the US$0.5–1 million range. This also means that a few large transactions (buy-outs) constituted the majority of investments (53 per cent), while the vast majority of the deals were much less than the average in terms of value.

The appearance of early stage investment was noticeable, being around 63 per cent of the transactions in 2000, and two-third, in 2001 by number

Table 10.2 Venture capital investment in Hungary, 1999–2001

	1999	2000	2001
Invested capital (million US$)	41	103	64
Number of transactions	12	47	28
Average investment (million US$)	3.4	2.2	2.3
Share of expansion capital, by no. of transactions (%)	100	36	11
Share of early-stage investment, by number of transactions (%)	0	64	71
Share of buy-outs, by number of transactions (%)	0	0	18
Share of high-technology investment, by value (%)	89	71	37
computer-related/electronics	6	21	11
information technology	12	4	12
telecommunication	0	17	4
biotechnology	71	16	10
Share of other investments (%)	11	29	63
Number of active VC firms	~20–22	~35	~30
Estimated number of VC professionals	500–600	600–800	800–900

Source: Based on *Hungarian Venture Capital Association Yearbook* (2000, 2001, 2002).

of transactions, compared to 1996–8, when it was zero. Expansion/development capital lost its importance. There were only 17 and three deals in 2000 and 2001, respectively. The appearance of buy-outs in 2001 (five) was associated with international movements. Out of the five buy-outs, two involved public-to-private selling and the other three were due to the reorganization of the foreign parent company. The unsatisfactory achievement and liquidity of the Budapest Stock Exchange (see later) and international mergers and acquisitions would have further effects on buy-outs.

The year 2000 was an exceptionally good: the invested capital exceeded US$100 million, the number of deals were higher than ever and the dominance of high-tech investments was a sign of the dynamically developing computer-related, telecommunication and biotechnology/pharmaceutical sectors. The number of active venture capital companies also hit a record level, however, the number of estimated venture capital professionals was still below a thousand.

Despite its small size, the Hungarian venture capital market continued its integration into the world market in the 1999–2001 period. In 2000–2001, around half of the deals were not only Hungary investments. In 2001, out of the 28 deals, nine were global and four were regional. The movement towards other countries in the region or the world was also a necessity: the Hungarian market in terms of both population size and income was too small for a dynamically growing company, especially in the high-tech sector. International expansion was one of the major reasons for Hungarian businesses searching for venture capital.

As a sign of the changing strategy, the increased importance of syndicate investment could be observed: in 2001, 13 transactions out of 28 included coinvestors. Partners were frequently involved in second and third-round investments that were also a new phenomenon in the Hungarian market.

Figure 10.1 presents the main characteristics of venture capital investment by industrial sectors in 2000–2001. The figure shows the sectoral changes of venture capital investment, that is the increasing importance of the technology/high-tech investment. Before 1998, only 7 per cent of the transactions happened in the technology sector, in year 2000, 70 per cent of the transactions were there. After the collapse of the new economy companies in the American stock markets, there was an adjustment in the Hungarian market too: traditional industries won back some space, up to 50 per cent from the new economy companies. The situation is not as satisfactory when we have a look at the transactions by value: in 2001, the technology sector constituted only 39 per cent of the investment. However there was an agreement amongst venture capitalists that Hungarian entrepreneurs are strong in software development, pharmaceuticals, biotechnology and some areas of laser application.

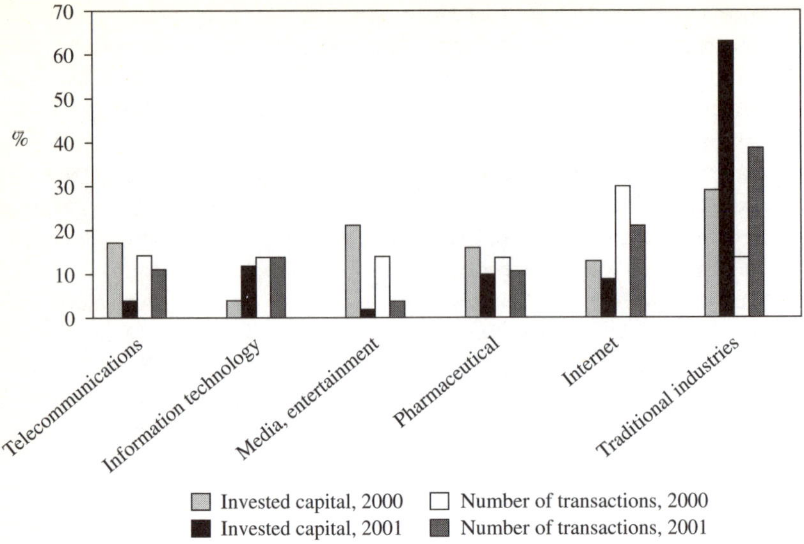

Source: Based on *Hungarian Venture Capital Association Yearbook* (2001–2002).

Figure 10.1 Venture capital investment in Hungary, by industry, 2000–2001

Since the venture capital market has only a few years' history, there were limited data about exits in Hungary. There is a common knowledge amongst experts that the progress of the VC industry is strongly associated with the development of the stock market (see, for example, Mani and Bartzokas, 2002). Stock markets are important for venture capitalists since they provide a perfect place for initial public offerings (IPO) for businesses that have gone through the expansion phase and for venture capitalists who consider selling their ownership share.

In general, Hungarian financial markets are relatively well developed, having a good infrastructure and a lot of further potential to develop. However the stock market shows a contradictory picture. Until now, the stock market has played only a minor role, when venture capitalists sold their shares. There were only four IPOs: North American Bus Industries (NABI) (machinery), introduced by the First Hungarian Fund, Synergon (operation integration systems), introduced by Advent, the Hungarian Private Equity Fund, and, in the Vienna stock market, E-PUB (software) introduced by Euroventures. After the successful premier in the Frankfurt Neuer Markt, Graphisoft (software development) was introduced to the BSE on 16 May 2001. Professional investors bought Recognita (software development) and Elender (Internet services) ('Jön a kockázati tőke', 2000).

Despite continuous acquisitions, NABI and Synergon have not used the stock market to raise capital. Synergon relied on retained earnings in its expansion to be a regional leader. NABI has used bank loans in its external growth financing in the USA and Great Britain.

The development of the Budapest Stock Exchange (BSE) was broken in 1998 and by 2001 and by now even the independent existence is being questioned. The capitalization of the BSE is small: annual turnover equals about the two days' turnover of the 30 largest companies traded in Frankfurt. Instead of new IPOs, companies are leaving the stock market and this is definitely not a good news for venture capitalists that plan to sell their shares on the BSE.

The limited use of the stock market as an exit route by the venture capitalist was also evidenced in Karsai (1999). Between 1995 and 1998, IPOs constituted only 7 per cent of the exits (4 per cent in terms of value), while trade sales amounted to 57 per cent (45 per cent in terms of value) and other methods (repurchase and refinancing) amounted to the remaining 36 per cent (51 per cent in terms of value). It is interesting that there was no involuntary exit in the same period – or that venture capitalists did not reveal them (Lemák, 2000).

In another empirical study about Hungarian venture capital investments and investors, Ludányi (2001b) claimed that the preferred exit method of most of the investors was trade sale. In contrast, state-owned venture capital companies favored selling their shares to fellow-owners, who were, for the most part, the entrepreneurs themselves (repurchase). In the case of trade sales, potential buyers were mainly domestic and foreign strategic investors who had had no ownership in the investee company before the exit.

In Hungary, there were only one fund of Euroventures that finished a full cycle of venture capital with an average 20 per cent yearly yield in 2000. As the company reported the majority of the portfolio of ten companies proved to be successful businesses. Five of these investee companies became a part of international or multinational companies, and two of them remained independent businesses under the new ownership. There were two failures: as András Geszti, Euroventure's Executive Manager, reported, all the invested money – around a million US$ – in Alfagrafix was a total loss (Csabai, 2002). Probably the one remaining investment was around the break-even point.

The limited number of exits in the 1999–2001 period were associated with the small size of the market, the increased appearance of early-stage investments and the unfavorable international conditions. In 2000, out of the ten exits, six were trade sales, two were IPOs and two involved repurchase and refinancing. There were no reported involuntary exits. In 2001, there were only four exits, but several write-offs.

As a consequence of the major problems of the new economy companies, investors in the Internet and high-tech sectors in Hungary tried to get rid of their investments. However, under unfavorable conditions, the only means of exit was to sell ownership back to the entrepreneur; otherwise the venture capitalist had to write off the whole invested capital. As a consequence, some Hungarian entrepreneurs had a very good chance to buy back the business fully or partially at a very good price (Elszállt a kedvük, 2002). While it looked a profitable deal for the entrepreneurs, who could gain a lot in the short run, decreased activity of future investors could hurt the development of the Hungarian new economy in the long run.

THE CONNECTION BETWEEN VENTURE CAPITAL AND INVESTEE COMPANIES: A CASE STUDY

In the following we present a case study of five venture capital companies and of their ten high-tech investments. Besides describing the characteristics of the venture capital and investee companies, the major aim of this section is to reveal the connection between them. Besides using publicly available data we carried out interviews with the representatives of the involved companies between March and June 2002. In October 2002, a follow-up telephone interview with the venture capital company representatives served to complete the survey.

Selection and Profile of the Interviewed Venture Capital Companies

As presented earlier, the Hungarian venture capital market is relatively small: in 2001, there were about 30 active venture capital funds. With the help of the HVCA, we approached seven and finally selected five companies that had been actively involved in high-tech investments. Table 10.3 shows the profile of the venture capital companies in our database. The table contains information that is publicly available.

Establishment

All of these funds were established after 1998 or later, up to 2000. The foundation was strongly associated with two events. First, because of the Russian crisis, some of the venture capital investments were lost completely, which made venture capitalists very careful with traditional industries. Second, the boom of the new economy sectors in the USA and the EU had a positive effect on new venture capital company and fund establishment.

Table 10.3 The profile of the venture capital companies/funds, 2001 (based only on publicly available data)

	3TS Venture Partners	Euroventures/ABN AMRO Capital	FastVentures	Hungarian Innovative Technology Fund	KFKI Investment
Year of establishment	1998	1998	2000	1999	1998
Format of business	Company	Ltd	Ltd	Ltd	Ltd
Ownership	Foreign	Foreign	Mixed	Foreign	Hungarian
Main owners/source of capital	3i, Sitra, SET, EBRD	ABN AMRO	Private investors	Hungarian–American Enterprise Fund	KFKI Számítástechnikai Rt, MAVA, HEP
Type of main owner(s)	Institutional (regional fund)	Bank	Private individuals	State (country fund)	Private
Capital under management/ invested capital	EURO66m./ confidential	Not limited/ EURO30m.	US$5.5m./ confidential	US$10m./US$1.6m.	Confidential/ US$14m.
Minimum/maximum investment	EURO0.5m. EURO5m.	EURO1m. EURO10m.	EURO0.1m. EURO0.75m.	US$0.05m. US$0.5m.	No data/US$1m.
The number of staff establishment, 2001/2	3/4	6/6	5/3	2/4	4/3
Number of portfolio companies, 2001/2	6/10	5/4	8/8	3/7	10/6
Stage(s) of financing	Seed, start-up, early phase, development	Start-up, early stage, buy-outs	Start-up, early stage	Seed, early stage	Seed, early stage
Industry preference(s)	IT, telecom, Internet	No specific preference	New economy	High-tech sectors	IT, telecom, health, fitness
Number of reported exits since foundation	0	1	3	0	4

Source: *Hungarian Venture Capital Association Yearbook* (2001), http://www.hvca.hu.

Four out of the five companies had a strong preference toward the new economy/high-tech sectors.

Business format

The venture capital companies preferred the limited liability company format that was easier to establish than a company limited by shares and where it was easier to maintain the close relationship with the owners.

Ownership – regional focus

Amongst the selected companies, all of the four main types of ownership forms could be found: institutional (regional and country funds), banks, private and state ownership. The two largest venture capital companies were in foreign hands, and only one of them had a domestic Hungarian majority in ownership. This data set represented the Hungarian venture capital market very well as regards ownership. The ownership form was closely related to the regional focus. The three foreign companies also had investments in other countries besides Hungary: 3TS Ventures had offices in Budapest, Prague and Warsaw. Moreover, as a part of ABN AMRO Capital, Euroventures belonged to an international network having several offices all around the world, in Chicago, London, Warsaw and Budapest, just to mention a few. Three out of ten investments of HITF were in other Central European countries. The remaining two companies, FastVentures and KFKI, had investments only in Hungary.

Size – capital under management

In terms of the size of the venture capital companies, there was marked variation: 3TS Venture and Euroventures/ABN AMRO Capital were large even by international standards, while two companies were much smaller, having US$5–10 million capital under management. We do not know the KFKI's capital, but it had already invested US$14 million, most of their available capital, so the capital under management was around US$15 million. The average size of the companies in our data set was much higher than a year before (Ludányi, 2001a).

Size – investment

The deals were in US$0.05–5 million, with the exception of Euroventures/ABN AMRO Capital, with deals up to EURO10 million. In the case of three of the five companies, this range was much lower than the average investment in Hungary before 1998 (more then US$1 million). However, it should be noted that venture capital companies were trying to invest close to the upper rather than the lower limit.

Size – staff

The average number of staff was four at the time of investment and also in 2002. However the venture capital companies that increased the number of companies in the portfolio employed more experts: FastVentures' staff decreased by two and KFKI's by one manager. The number of staff at the venture capital companies was not known exactly because of the varied participation of the owners. If there was a need (because of new companies, crisis, expansion and so on) – some of the founders spent more time at the venture capital company, while in the case of shrinking portfolios and fewer tasks they were less active.

Out of the five companies, Euroventures had the largest staff but the smallest portfolio. The reason for this was that the experts not only handle the Euroventures funds but also participate in other equity businesses of the main owner, ABN AMRO. It was estimated that two or three people dealt with venture capital investments.

The portfolio – industry and stage preferences

The number of portfolio companies varied from three to10, averaging at 6.8 in 2000 and 7 in 2002. While 3TS and HITF increased the number of investee companies, KFKI's porfolio became smaller. FastVentures replaced some of the companies with others, but its portfolio was the same in terms of the number of companies. Euroventures' portfolio decreased by one, it will be launching new investments presently.

We should note that the average investment was much lower at the companies that had a larger number of firms in their portfolio. There was also a potential overlap in the portfolios since venture capital companies in high-tech industries frequently prefer coinvestments in order to decrease and spread risk. Similar to other high-tech/new economy investors, our companies preferred early stage and/or seed capital investments and, with one exception, had a strong preference for the new economy sectors.

All of the five venture capital companies preferred minority ownership, one of them favoured qualified minority (ownership over 33 per cent). All of the companies used capital stock as a main instrument of financing. Minority rights were protected by co-sale (mostly tag along) rights and shares that have specific rights (preferred shares). Two of the companies refused to use ownership loans, and three of them relied on transferable bonds. It was commonly believed that loans were not a proper way of financing firms by venture capitalists. Stock option plans were used by one company, which was not a surprise considering the situation of the Hungarian stock market.

Exits – exit strategies

Regarding exits, all of the five companies would prefer the stock market (IPO) when they wanted to realize their investment. However the situation of the Budapest Stock Exchange (BSE) did not make IPOs attractive. As an alternative, one of the companies tried to sell ownership back to the entrepreneur (repurchase). The others preferred professional strategic investors (trade sales) or postponed or cancelled new investment if they were not sure of the exit conditions. Actual data on exits showed that there were two companies out of the five with a larger number of exits: FastVentures had three (and one partial) and KFKI had four exits. We have no data on the exit method but it probably did not differ from what we stated previously: trade sales and repurchase dominated, while IPO was zero. At the same time period, two other venture capital companies – 3TS and HITF – considerably increased the number of portfolio companies. Therefore the total number of investee companies did not change in the data set between 2000 and 2002.

Strategy – change

All of the venture capital companies were established under favourable conditions in the new economy. When there was a boom in the high-tech industry, venture capitalists poured money into the new sector companies much less carefully than previously. However this was only partially true in Hungary, where foreign investors were traditionally more risk-averse than in other developed countries. Investing in a risky sector in an early phase in a small business for a longer time means much more risk than developing a traditional, established, large company for a shorter time. Therefore venture capitalists tried to decrease the risk by buying only minority ownership and working together with other venture capital companies. Instead of investing one large amount they preferred to give money in smaller portions, but more than once if the investee company fulfilled the expectations.

After the collapse of the new economy companies, three out of the five venture capital companies changed their strategy: 3TS balanced its portfolio by moving towards larger businesses that were not in the early stage of development. FastVentures refused to provide seed capital and also moved towards more mature businesses. They also required owners and managers more committed to the development of the investee company. KFKI also moved towards the development phase of investment. Euroventures and HITF did not change their strategy. Ferenc Berszán (HITF) stated that the company had traditionally had a conservative investment policy. Thanks to careful selection it did not have major failures and losses. Euroventures was a little different from the other venture capital companies since initially it focused more on large businesses in the development phase.

Deal flows

There were marked differences in the volume of deal flows. If we define serious interest as submitting the business plan, then the deal flow ranged from around 100 to 25. Two companies had about 100, one about 80, one 50 and one 25 offers in a year. The average number of offers was 70. Some of the venture capital companies searched more actively than the others. The search activity depended on several conditions: we could see that companies were more active if they were younger, had a larger portfolio, wanted to increase the portfolio, and specialized in certain sectors (such as biotechnology).

It is worth noting, that the number of deal flows was not associated with the size of the capital of the company. There were marked variations amongst the examined companies regarding the deal flow strategy. All of the companies agreed that 'word of mouth' played an important part as a marketing tool. Moreover the company managers also had a wide personal network. As a general promotion tool, everybody tried to be in the news, mostly in the professional dailies and journals. Exploitation of TV and radio news and major conferences, such as the HVCA yearly conference, as well as certain large investments, was much less frequent. All of the five companies relied on the owners' network. Venture capitalists were frequently involved in conferences and workshops. Two of the companies identified conferences as the most important place for meeting potential investee company representatives. HVCA membershipship also had a positive effect on deal flows. Businessmen's meetings were mentioned, but only by one company. It was a surprise that one venture capital company did not have a web page. Two companies used outside agents, and First Tuesday events as an important meeting place was mentioned by two companies. There was only one company, HITF, that used all of the activities, including the personal search for new deals. As they stated, competition had been increasing, which forced them to use a variety of tools in seeking new deals. Moreover in certain sectors (biotechnology was emphasized) personal search was inevitable.

Rate of return

There were variations in the expected rate of return. Three of the five companies expected a 25–30 per cent rate of return, one 40–50 per cent and one sought more then 50 per cent returns on a yearly basis. These returns applied to US dollars or EUROs. It should be noted that there was no exact rate of return, venture capitalists considered various conditions, including industry, risk and market conditions, when they made the final decision about investment. Past internal rates of return were within the 25–30 per cent range, despite some lost investments.

The Selection and Profile of the Investee Companies in the Data Set

The estimated number of investee companies in the new economy segment was around 50, most of which received venture capital more then once. Of these 50 companies, ten were selected for our database (see Table 10.4). The choice of the investee companies in the data set was a mixed procedure of media news and venture capital company suggestions, limited by the requirement that they should represent at least five high-tech sectors.

Because of business confidentiality we do not give the name of the company and its connection to the venture capital firm. Despite guaranteed privacy, some of the sensitive data were not revealed exactly. However Table 10.4 makes it possible to have an inside view on Hungarian investments. Unlike the case of venture capital companies, there has not been any research regarding investee firms in Hungary, so this is the first time that Hungarian investee companies' profiles have been analysed. Since most of these data are confidential, there are some missing elements in Table 10.4.

It was agreed initially that the ten selected investee companies must represent at least five high-tech sectors. It was not a surprise that three out of the ten companies were in the telecommunications sector and two of them in the pharmaceutical sector. The electrical equipment sector was represented by two firms, with one company belonging to each of the non-electrical equipment, the computing and office equipment and the chemical sectors. Some companies had mixed activities, which is why sometimes we have two sectors in one box.

Capitalization of the firms – where data are available – shows marked variations. The highest figure was around US$100 million, for a Hungarian subsidiary of an international company. (The venture capital investment was used to support the research activity in Hungary.) The lowest figure for company capitalisation was US$0.76 million, for a medium-size company in Hungary.

In all of the cases, venture capital companies gained minority ownership, ranging from 2 to 40 per cent. The average amount of investment was about US$0.89 (EURO1m.), which was less than half the average venture capital investment in the years 2000 (average investment US$2.1m.) to 2001 (average investment US$2.3m.). It came as no surprise that the largest investment went to the most capital-intensive telecommunications companies (US$1.3m., EURO2.1m., EURO2.65m). The other six known investments were below US$1 million, and probably the remaining company investment was also well below that figure. Without telecommunications investment, the average money put into a company was about US$372000, less than 20 per cent of the average investment in 2001.

The relatively low amount of venture capital investments in the high-tech sector could be explained by several factors:

- investors were very careful, as they want to risk a low amount of money in one company;
- since most of the investments were in the early phase they required relatively less money;
- follow-on investment increasing, if the initial project met with the expectations, then further investments would took place;
- the strategy to share risk by providing syndicate or co-investments became more common.

It is interesting to analyse why companies/entrepreneurs turned to venture capitalists and what they expected from them. As seen in Table 10.4, two companies wanted to continue research that related to the development of new drugs, medicine or molecules. Both companies in this area belonged to an international parent and have a research network. The search for venture capital was associated with the high expense and high risk of failure of this type of research, as well as the limited money for R&D in Hungary. Another pharmaceutical firm that already had the product was aiming to enter the market (seed capital). Company 8 had no clear strategy about the product and the company's future and was looking for help. In this case the most important problem was the market.

In all the other six cases – independently of whether the investments were in the early or in development phases – the investee companies were looking to expand their business domestically, but mainly internationally or regionally. So, besides the money, they expected the venture capitalist to provide help to search for markets or provide actual connections to other buyers. The internationalization of high-tech business is vital in a small, developing country like Hungary. On the one hand, the domestic market and consequently the number of potential buyers are very small, while on the other hand the demand for high tech products is also limited because of the development stage of the country. Moreover prompt exploitation of existing opportunities in the rapidly expanding high tech business is inevitable. Earlier successful Hungarian high-tech companies like Recognita and Synergon used the venture capital received to set up branch and/or selling offices in the USA and the European Union.

It is well known in the literature (for example, Reynolds *et al.*, 2001) that not only formal venture capital but also informal business angel financing can be very important for high-tech businesses. It is common in the USA for a company to start with a relatively low amount of angel money and then continue with a larger venture capital investment. In Hungary, business

Table 10.4 The basic profile of the investee companies in the database

	Year of establishment	Profile/technology sector	Total capital	Ownership share of VC company	Invested amount	Aim of investment	Number of VC investment/ angel finance	Number of co-investors
1.	1998	Internet/telecommunications	Not revealed	Not revealed/ minority	EURO2.1m.	Regional expansion	2/yes	3
2.	1993	e-business, telecommunication	~EURO7m.	Minority (~40 per cent)	EURO2.65m.	International expansion	1/no	1
3.	1997	Medical drug research/ pharmaceutical, chemical	~US$35m.	Minority (2–5%)	US$0.4m.[1]	Research	4/yes	8
4.	1999	Chemical, biotechnology	Over US$100m.[2]	Not revealed/ minority	US$0.46m.	Research	3/no	6
5.	1995	Laser technology/ non-electrical equipment	Over US$15m.	Minority[3]	US$0.7m. US$6m.[4]	Development, international expansion	2/no	4
6.	1999	Biotechnology/pharmaceutical	Not revealed	Minority	US$0.2m.	Market entry	1/yes	1
7.[5]	1998	Telecommunications	US$12m.	Minority (12%)	US$1.3m.	Domestic expansion	2/no	1
8.	2000	Computing and office equipment	Not revealed	Minority	US$0.1m.	Strategy formulation	1/no	1
9.	1993	Electrical equipment, telecommunications	US$2m.	Minority	Not revealed	International expansion	2/yes	1

10[6]	1989	Electrical equipment	US$0.76m.	Minority (21%)	US$0.16	International expansion	1/no	1

Notes:

1. The total investment was US$11 million, which means altogether a 32 per cent venture capital share in the company.
2. The Hungarian company is a subsidiary of a German parent company; the total capital belongs to the whole company.
3. The venture capital companies altogether have a 65 per cent ownership share in the company.
4. The second-round investment that was provided by a syndicate of four was worth US$6 million and a 65 per cent ownership share altogether. Only the first investment is included in the analysis.
5. The data are for the first investment; there are no data available regarding the second investment by the same company.
6. The company has been sold to a professional investor (trade sale).

angels represent only 2.2 per cent of the total population, that is, less than the world average and much less than in the leading countries, New Zealand and the USA (with around 6 per cent). However, the importance of business angels can be seen in Hungary: four out of the ten companies had received angel money before the venture capital participation.

The Connection between the Venture Capitalist and the Investee Company: Controlling and Participating

A distinguishing feature of the venture capital companies is to provide help to the investee company as well as money. The reason for the additional help is twofold: first, it is to gain strong control over the company – much stronger than in the case of bank loans; second, most of the companies are not properly managed. Leadership, management, marketing, financial planning, reporting, strategy focus and so on are frequently missing. Moreover, most of the time, the investee companies want to expand the businesses. This means that the previous management techniques, even if they used to be perfect, cannot be applied any more. The company has to deal with changes in sales, personnel, organization, buyers/sellers connections and so on at the same time. Venture capitalists provide help to find the solution to manage growth. It can be believed that in a transitional country, like Hungary having only 12 years history in market-economy practice, entrepreneurs and potential venture capital investee companies have less experience in management and leadership than in other countries having longer history in the market economy.

This fact was reinforced by venture capitalists: the lack of proper managerial skills is one of the main limitations on further venture capital investments in Hungary. As a consequence of this less experienced management, venture capitalists may follow two practices. They select the investee companies much more carefully than in other countries and/or they engage more actively in the management; the second solution, of more active participation, could be more expensive in terms of costs that would, however, increase the expected rate of return substantially. The risk is even greater if the company is in the early phase and the owner(s) have no managerial and business experience at all.

We can find both of the selection strategies amongst our venture capital companies. Two of the companies focused more on the selection criteria and made no distinction between managing a company in the early phase or one in the development phase. However, they chose companies that have a time-tested management team. All the other venture capital companies differentiated between managing research, early phase or developing companies. The earlier the investee company in the life cycle the more time was necessary for the venture capitalist to participate in the management. While two venture capital companies required quarterly reports, three expected monthly reports.

When we asked the venture companies to estimate how much time they spent at the investee company, this proved to be difficult. There was agreement that one venture capital manager can manage two to four projects. Besides Euroventures, which was less active in management, three out of the other four venture capital companies spent about a day a week, and one spent half a day in a week at an investee company. In the early phase of the investment, meetings between the venture capitalist and the investee company could be daily. HITF requested the company to send daily cash-flow details, 3TS participated in negotiations, FastVentures focused on avoiding crisis situations. In other cases, the venture capital company helped the investee company to set up the financial reporting system. The venture capitalist's help was vital in the early phase when the investee company had no resources to pay a professional employee.

We have already mentioned that Hungarian entrepreneurs possess fewer business skills than entrepreneurs in more developed countries. Three of the five venture capital companies complained more or less about the lack of vital management skills and insufficient commitment to the project. This was probably the reason why these venture capital companies began to move towards later phase and larger investments, where the management team had already had some experience and proof of survival capacity. More attention was given to the commitment of the entrepreneur both financially and non-financially.

Table 10.5 provides the necessary information about the connection of the venture capital and the investee companies on a company basis. The first five categories – strategy formulation, financial control, investment control, leadership control and production/research control – are the most important tools for the venture capitalists to monitor the investee company. The other categories describe the type of help that the venture capital company provides to the investee firm.

By examining the tools of control, Table 10.5 reveals that venture capitalists practice both financial and leadership control. The close financial control, including the regular follow up of the investee company's financial performance (sales, costs, financial ratios and so on) and participation in the additional investments were common. However regularity did not mean a daily involvement: most frequently a monthly report was required. There were only two companies where venture capitalists did not participate in financial/strategic planning. The reason was that these companies were in the research phase, and there were no sales at that time. However, as a substitute, venture capitalists had a close look at the research procedure. In all

Table 10.5 Characteristics of the connections between the venture capital and the investee companies

Categories/companies	A	B	C	D	E	F	G	H	J	K
Financial/strategic planning	++	+	++	+	++	++	–	++	–	++
Controlling finance	++	++	++	++	++	++	++	++	++	++
Controlling additional investment(s)	+	+	++	+	++	++	–	++	–	+
Controlling leadership	++	++	++	++	++	++	++	++	–	++
Following production/research	–	–	–	–	–	–	++	–	++	–
Help to get additional financial resources/ bank loan(s)	+	+	+	+	+	+	+	+	+	+
Participating in leadership formulation	+	+	+	+	+	+	–	+	–	+
Help in recruitment of key personnel	–	+	+	–	+	++	–	+	–	+
Help to get access to new technology	–	+	++	–	+	–	–	+	–	+
Help in connection/network building	++	++	++	++	++	++	++	++	–	++
Help to get into international market	++	+	+	+	++	+	++	++	++	++
Participating in marketing strategy formulation	++	–	–	–	–	–	–	+	–	–

Notes: The strength of the connection is indicated by '+' and '–' signs: '++' = strong participation/help, '+' = occasional (semi-strong) participation/help, '–' = no participation/help.

the other eight cases, investors were not participating in the production process.

Authority over the leadership (management) is mainly practiced via the board of directors or regular assembly meetings. The magnitude of the authority, however, depends on the number of delegates of the venture capitalists on the board. Since in most of the cases in our data set venture capitalists had minority shares this meant that the entrepreneur or original owners had a majority on the board. Despite the limited control over the management, personal conflicts were rare because venture capitalists had a close look at the management in the due diligence phase of the selection procedure. If they had any doubts about the personality, capability or skills of the management/owners, they would not invest in the business. This role of the venture capitalist was reinforced by the relatively weak or occasional participation in leadership formulation.

While control over the investee company serves the interest of the venture capitalist, the additional help provided by the investor can be vital from the viewpoint of the company. There were two areas where the companies in our survey received considerable help from the venture capitalists. In the high-tech sectors, network building and close personal contacts with other similar companies or potential buyers were inevitable. Hungarian venture capitalists helped the investee firms to find the proper connection. As noted earlier, entering the foreign markets was overwhelmingly the most important reason for the companies' search for venture capital. According to Table 10.5, it appears that venture capitalists fulfilled the initial expectations of the investee firms and provided major help to get access to, and widen, foreign relations. By far the most important foreign relations were with the USA and the European Union, mainly Germany. There was only one investee company that complained about the insufficient help of the venture capitalist in foreign relation building.

It can happen that there is a need to obtain additional financial resources, mainly bank loans. In these cases, venture capitalists provided limited help and most of the time they refused to be a guarantor of the bank loan. Sometimes they participated in finding key personnel. In the case of company F, the venture capitalist carried out an extensive international search to find the proper expert. In half of the cases the venture capitalist helped the companies to develop or to find the proper technology. However, technology was only vital at one telecommunications company. This also shows that the investee companies had possessed the necessary technology at the time of investment.

Marketing is probably one of the weakest points of Hungarian businesses. While there are excellent ideas and creative people, marketization of innovations is far from satisfactory. We believed initially that venture capitalist would pay special attention to the marketing strategy of the investee company. However this was true only in one case and there was another case where some marketing strategy participation of the venture capitalist could be seen. The reason for this behaviour was probably associated with the fact that venture capital managers were mainly financial experts.

CONCLUSION

Since its existence, venture capital industry has been in heaven and hell. The enormous boom in the American new economy had a crowding-out effect on the world. While Europe followed the USA with a little lag, the influence on Hungary arrived even later, in 2000. This is the main reason why the collapse of the new economy companies in the USA did not have a major effect on Hungary: the whole new economy constitutes only a small part of the Hungarian economy.

The examined venture capital companies were established in the late 1990s to early 2000s with the aim of making high-tech investments in Hungary. At the same time, existing venture capital companies turned more to new economy and early stage investment, following worldwide tendencies in VC investments. Following the exceptionally good year of 2000, venture capital investment decreased by 40 per cent in 2001, but, the high-tech sector retained its dominant position in VC investment.

It was shown in the case study that venture capital companies became more careful in selecting investee firms, and followed their performance in a more consistent way. Despite difficulties in comparing developed country venture capital companies' behaviour with Hungarian ones, some variations can be identified. Even in hard times, some venture capitalists increased the number of portfolio companies while others tried to get rid of risky, unsuccessful investments. Some of the companies focus more on selecting a firm that has a proven management, others are more willing to teach the investee company managers and owners how to manage growth. Lack of managerial knowledge and commitment of the entrepreneur have been the main limitations on further venture capital investments.

Because of the problems in the small, local Budapest Stock Exchange, IPO, as a desirable exit route, is not possible. Therefore trade sales played the primary role in the exit procedure. In some cases an imposed 'pushed' repurchase technique was used, but most of the venture capitalists waited for better exit conditions.

Examining the available capital in the region, it can be seen that there is still plenty of money, but there is a lack of proper investment opportunities, as venture capitalists stated. The latest report of Deloittle & Touche (2002) shows that Hungary is the most promising nation in the Central Eastern European region, having high-tech, fast-growing companies that could be the potential target of venture capitalists. Owing to a high level of entrepreneurial activity, the number of innovative start-ups may increase too. It is also promising that the Hungarian economy is growing at a rate double that of the European Union. The expected accession to the EU could also have a positive effect on the development of the Hungarian venture capital market.

Hungary has done the most painful part of transition. By 2002, the country and the Hungarian venture capital market were firmly integrated into the globalized world economy. As the importance of global and regional deals has been increasing, Hungary has a good chance of getting rid of the disadvantages of the small size and limited domestic income and of becoming the regional centre of venture capital.

ACKNOWLEDGMENTS

The authors would like to thank the Hungarian Venture Capital and Private Equity Association, its secretary István Lakos, and the venture capital representatives, Ferenc Berszán, Péter Geszti, László Hradszky, Péter Fodor, Balázs Bedő and Benedek Lőrincz. Also thanks to Judit Karsai for helpful comments.

REFERENCES

Acs, Z., L. Szerb, J. Ulbert and A. Varga (2002), 'GEM 2001 Magyarország – Vállalkozások Magyarországon globális összehasonlításban' (GEM2001 Hungary Report), University of Pécs.

B.I. (2001), 'Kockázatató töke', *Cégvezetés*, April, 122–5.

Csabai, K. (2002), 'Kiszálláscsinálók', *Heti világgazdaság*, **2**, February, 101–3.

Deloitte & Touche (2002) 'Fast 50 companies in Central Europe 2002 technology industry survey' (http://www.deloittece.com/fast50/result2002.php 3.bak).

Elszállt a kedvük (2002), 'Heti világgazdaság melléklet Kalauz az internethez', *Augusztus*, 6–9.

Halaska, G. and T. Kovács (1999), 'Színtévesztők renénysége', *Figyelő*, 25 November–1 December, 42–9.

Hungarian Venture Capital Association Yearbook (1999), (2000), (2001), (2002) (http://www2.hvca.hu/hun/publications.htm).

'Jön a kockázati tőke', *Világgazdaság*, 20 January 2000, pp.13–14.

Kállay, L. (2000), 'Mikrohitelezés piaci alapon', *Közgazdasági Szemle*, **47**(1), 41–63.

Karsai, J. (1997), 'A kockázati tőke lehetőségei a kis- és középvállalatok finanszírozásában', *Közgazdasági Szemle*, **44**(2), 165–74.

Karsai, J. (1999), 'A megfontoltan kockáztató tőkések', *Közgazdasági Szemle*, **46**(9), 789–812.

Karsai, J. (2000), 'A koxkázati tőke szerepe a technológiai alapú induló vállalkozások finanszírozásában', in *Felzárkózás és EU csatlakozás, A VII. Ipar- és Vállalatgazdasági Konferencia előadásai*, Budapest: MTA Ipar- és Vállalatgazdasági Bizottsága, pp.324–30.

Karsai, J. (2002), 'Closing the Equity Gap', The Funding of Venture Capital in Hungary: Issues for Public Policy', paper presented at an international conference on 'Evolution of Institutions and the Knowledge Economy', 3–5 October, Debrecen, Hungary.

Karsai, J. and A. Rácz (2000), 'Kockázati tőkések', *Cégvezetés*, June, 116–21.

Lemák, G. (2000), 'Kockázatitőke-társaságok Magyarországon', *Bankszemle*, **4**, 18–43.

Ludányi, A. (2001a), 'A tőkeerő és az alapítói háttér hatása a kockázatitőke-szervezetek befektetési magatartására – I', *Közgazdasági Szemle*, **48**(7–8), 659–72.

Ludányi, A. (2001b), 'A tőkeerő és az alapítói háttér hatása a kockázatitőke-szervezetek befektetési magatartására – II', *Közgazdasági Szemle*, **48**(9), 779–97.

Mani, S. and A. Bartzokas (2002), 'Institutional Support for Investment in New Technologies: The Role of Venture Capital Institutions in Developing Countries', *United Nations University/INTECH discussion papers*, Maastricht, the Netherlands.

Reynolds, P., M. Hay, W. Bygrave, S. Camp and E. Autio (2001), *Global Entrepreneurship Monitor 2001 Executive Report*, London: London Business School.

Szerb, L. and J. Ulbert (2002), 'Entrepreneurial growth and the role of venture capital in Hungary' in Attila Varga László Szerb (eds), *Innovation, Entrepreneurship and Regional Economic Development*, Faculty of Business and Economics Pécs: University of Pécs, 122–46.

Index